Springer Climate

Series Editor

John Dodson, Institute of Earth Environment, Chinese Academy of Sciences, Xian, China

Springer Climate is an interdisciplinary book series dedicated to climate research. This includes climatology, climate change impacts, climate change management, climate change policy, regional climate studies, climate monitoring and modeling, palaeoclimatology etc. The series publishes high quality research for scientists, researchers, students and policy makers. An author/editor questionnaire, instructions for authors and a book proposal form can be obtained from the Publishing Editor.

Now indexed in Scopus® !

Kai-Uwe Schrogl · Christina Giannopapa ·
Ntorina Antoni
Editors

The Politics of Geoengineering

Perspectives from the Social Sciences

Springer

Editors
Kai-Uwe Schrogl
Institute of Political Science
University of Tübingen
Tübingen, Germany

Christina Giannopapa
Department of Industrial Engineering
and Innovation Sciences
Eindhoven University of Technology
Eindhoven, The Netherlands

Ntorina Antoni
Department of Industrial Engineering
and Innovation Sciences
Eindhoven University of Technology
Eindhoven, The Netherlands

ISSN 2352-0698 ISSN 2352-0701 (electronic)
Springer Climate
ISBN 978-3-032-10094-8 ISBN 978-3-032-10095-5 (eBook)
https://doi.org/10.1007/978-3-032-10095-5

This Springer imprint is published by the registered company Springer Nature Switzerland AG
The registered company address is: Gewerbestrasse 11, 6330 Cham, Switzerland

Foreword

Geoengineering remains a loaded term and as such a source of debate in the scientific community. For this reason, it is essential that research continuous to be done on this concept whereby different aspects are addressed, from the ethical and legal side to potential risks, opportunities, and governance related matters.

Geoengineering origins trace back to early weather modification experiments in the mid-20th century, when military and scientific communities explored the potential to influence rainfall, hurricanes, and other atmospheric conditions. Yet it was not until the late 20th and early 21st centuries where it became a potential, albeit controversial, response to accelerating climate change. It emerged as an element in global effort to combat climate change, offering alternative options to mitigate greenhouse gas emissions and manage global warming.

This volume, *The Politics of Geoengineering: Perspectives from the Social Sciences*, provides a layered understanding of geoengineering, beginning with foundational questions and extending toward institutional, legal, and public engagement dimensions. Across its chapters, the book offers a progression from conceptual framing to governance challenges and normative considerations.

The opening chapter sets the analytical foundation by examining how geoengineering is framed within academic, political, and media discourses. It highlights the importance of framing effects in shaping perceptions, research agendas, and policy priorities. This chapter establishes the conceptual tools necessary to interpret the political dynamics that follow.

Subsequent chapters explore the role of international institutions and legal regimes in responding to the governance challenges posed by geoengineering. Particular attention is given to the fragmented nature of existing international environmental law, the limitations of current legal instruments, and the implications for anticipatory regulation. These contributions demonstrate how existing frameworks, such as the London Protocol or the Convention on Biological Diversity, interact with the emerging geoengineering agenda.

Further chapters examine the governance landscape through the lens of scientific diplomacy, geopolitical asymmetries, and the role of epistemic communities. Case studies illustrate how national and transnational institutions are navigating the uncertain terrain of geoengineering oversight. The analysis underscores the tension between scientific expertise and democratic legitimacy in decision-making processes.

The final chapters turn toward societal dimensions, addressing questions of legitimacy, justice, and public engagement. Here, the book considers the role of civil society, the conditions for inclusive dialogue, and the need for transparency in both research and governance. These chapters emphasize that geoengineering is not solely a technical or regulatory matter, but one embedded in broader societal and ethical contexts.

Taken together, the chapters offer a coherent, multi-dimensional analysis that reflects the complexity of governing emerging climate technologies. By drawing on a range of social science disciplines and methodologies, the volume contributes to the growing body of scholarship seeking to understand and shape the governance of geoengineering in a way that is scientifically informed, ethically grounded, and publicly accountable.

Eindhoven, The Netherlands

Robert-Jan Smits
President Emeritus of the Eindhoven University of Technology
Former Director-General for Research and Innovation, European Commission

Contents

Introducing GE

Introduction: Turning Earth's Thermostat Down—But how? The Politics of Geoengineering. Perspectives from the Social Sciences 3
Kai-Uwe Schrogl, Christina Giannopapa, and Ntorina Antoni

SWOT Analysis for Carbon Dioxide Removal (CDR) 39
Valeria Pinna, Leonora Burnik, and Laure-Marine Vioujard

SWOT Analysis for Solar Radiation Modification (SRM) 59
Valeria Pinna, Leonora Burnik, and Laure-Marine Vioujard

International Political and Legal Aspects

Governing Carbon Dioxide Removal and Solar Radiation Modification 81
Joshua Horton

Asia-Pacific Perspectives on Solar Geoengineering 95
Masahiro Sugiyama

Who Will Bell the Cat? African Union, Geoengineering and the Potential Role of Human Rights 109
Ademola Oluborode Jegede

Reflections on Geoengineering and Biodiversity 125
Perses Bilimoria

Geoengineering and International Law 133
Alla Pozdnakova

Solar Geoengineering and the Global Commons—What Role for Ecocide Law? 147
Anna Faye Maddrick

Geoengineering and Outer Space: Policy and Legal Considerations 161
Anthi Koskina and Manolis Plionis

Societal and Public Aspects

Geoengineering and Public Value: Toward a Global Climate Sensibility 175
Rick Wylie

Public Engagement with Geoengineering: Emerging Technologies, Experimentation and the Climate Crisis 191
António Carvalho

GE Extended: Terraforming on Celestial Bodies 203
Chris Pak

About the Editors

Prof. Dr. Kai-Uwe Schrogl is Special Advisor for Political Affairs at the European Space Agency (ESA) and Honorary Professor for international technology policy at Tübingen University, where he since a number of years also teaches courses on geoengineering. Previously he was seconded from ESA to the German Federal Ministry for Economic Affairs and Energy in Berlin to support the preparation of the German Presidency of the Council of the European Union in the second half of 2020. Until 2019, he was the Chief Strategy Officer of ESA (Headquarters in Paris, France), in which capacity he introduced geoengineering into the study activities of the agency. From 2007 to 2011 he was the Director of the European Space Policy Institute (ESPI) in Vienna, Austria, the leading European think tank for space policy. Prior to this, he was the Head of the Corporate Development and External Relations Department in the German Aerospace Center (DLR) in Cologne, Germany. He has been a delegate to numerous international forums and has served from 2014 to 2016 as chairman of the Legal Subcommittee of the United Nations Committee on the Peaceful Uses of Outer Space (UNCOPUOS) the highest body for space law making, comprising more than 70 Member States at that time. He also was chairman of various European and global committees (ESA International Relations Committee and two plenary working groups of the UNCOPUOS Legal Subcommittee, the one on the launching State and the other on the registration practice, both leading to UN General Assembly Resolutions. Recently he co-chaired the Global Council on Space in the World Economic Forum WEF. Kai-Uwe Schrogl is the President of the International Institute of Space Law, the professional association of space law experts from more than 50 countries. Kai-Uwe Schrogl has written or co-edited more than 20 books and more than 150 articles in the fields of technology policy, space policy and law as well as telecommunications policy. He sits on editorial boards of various international journals in the field of space policy and law in Europe and Asia.

Dr. Christina Giannopapa is the Head of the Office of the Executive Director of the European Union Agency for the Space Programme (EUSPA). Prior, she was advisor to the Secretary General of Telecommunications and Post of the Ministry of Digital

Governance in Greece, seconded from the European Space Agency (ESA). In ESA, she held various posts including the Head of Political Affairs Office of the Director General. From 2010 to 2012, she worked at the European Space Policy Institute (ESPI), seconded from ESA, where she had been managing the secretariat of the European Interparliamentary Space Conference (EISC) and led studies on innovation, Galileo, Copernicus and Africa. She also worked briefly in DG Research at the European Commission. Prior to joining ESA, she gained further experience as a consultant to high-tech industries in research and technology development. She held various academic positions, notably at the Eindhoven University of Technology and the University of London. She has received 14 academic scholarships and awards and has more than 80 publications in peer-reviewed journals, books and conferences. She has co-edited the Springer "Handbook on space security". Dr. Giannopapa is an elected member of the International Institute of Space Law (IISL) and the International Academy of Astronautics (IAA). She is the Vice Chair of the Committee on Space Security of the International Astronautical Federation (IAF) and prior was the Chair of the Committee for Liaison with International Organisations and Developing Nations (CLIODN). She is also the Chair of the Board of Women in Aerospace Europe (WIA-E). Dr. Giannopapa holds a Ph.D. in Engineering and Applied Mathematics; an MEng in Manufacturing Systems Engineering and Mechatronics; and an MBA in International Management from the University of London. Additionally, she holds an assistant professor position in the Department of Industrial Engineering & Innovation Sciences at the Eindhoven University of Technology.

Dr. Ntorina Antoni is a Postdoctoral Researcher specializing in innovation ecosystems at Eindhoven University of Technology. Her research focuses on strategic decision-making processes, stakeholder management, and strategy innovation within large high-tech organizations. Her Ph.D. thesis titled 'Managing Strategy Innovation: Co-Navigating the Strategy Perimeter in New Strategic Initiatives' was conducted with the European Space Agency as case study. Currently she coordinates the COOPERATE project which fosters intra- and interregional ecosystems for knowledge production and promotes regional best practices for innovation ecosystems across Europe. Dr. Antoni holds a Law degree from the School of Law of the National and Kapodistrian University of Athens, Greece. She also holds a Master of Laws (LL.M.) in International and European Law with cum laude from Tilburg University, and an Advanced Master of Laws (LL.M.) in Air and Space Law with distinction from Leiden University. Prior to joining academia, Dr. Antoni gained extensive experience in law, policy and strategy in the aerospace industry. From 2016 to 2017, she worked as strategy and policy analyst at the European Space Agency Headquarters in Paris, France, where she was involved in the analysis of regulations, policies, and strategies of various entities within the member states such as agencies, ministries, national parliaments as well as the European parliament and international institutions. From 2014 to 2016, Dr. Antoni worked as the legal counsel for Swiss Space Systems, a Swiss-based company that aimed to offer orbital launches of small satellites and manned sub-orbital spaceflights. In this role, she was responsible for overseeing the legal and regulatory aspects of aviation and aerospace projects, including aircraft purchases, certification of parabolic flights and small satellite launches. While working in the industry, Dr. Antoni also obtained her license as Attorney-at-Law in Athens, Greece, where she practiced as an attorney for several years.

Introducing GE

Introduction: Turning Earth's Thermostat Down—But how? The Politics of Geoengineering. Perspectives from the Social Sciences

Kai-Uwe Schrogl, Christina Giannopapa, and Ntorina Antoni

Abstract This introductory chapter examines the politics of geoengineering, analyzing how deliberate large-scale climate interventions have evolved from theoretical concepts to contested policy realities. Through a multidisciplinary framework integrating political science, economics, and innovation science, we trace geoengineering's trajectory from early 20th-century proposals to contemporary governance challenges. Our analysis identifies six critical analytical dimensions—environmental, political, economic, legal, societal, and justice-based—that shape geoengineering development. Moreover, it reveals an increasingly polarized landscape where technological capabilities advance faster than governance frameworks, creating a "risk versus risk" scenario between climate inaction and potentially disruptive interventions. Drawing on national policy developments, international regulatory efforts, and emerging funding patterns, we demonstrate that geoengineering now requires urgent

Important Note This book has a knowledge cutoff date of January 1, 2025. Consequently, several significant developments that occurred after this date are not addressed, though their importance is acknowledged. Most notably, the International Court of Justice (ICJ) delivered a landmark unanimous advisory opinion on July 23, 2025, clarifying that states have legally binding obligations under international law to protect the climate system from greenhouse gas emissions and to prevent global warming from exceeding 1.5°C. The ICJ ruling represents a watershed moment in international climate law, providing unprecedented legal clarity on state obligations and potential consequences for climate inaction. Similarly, any policy changes or implications resulting from the Trump administration, which began its second term on January 20, 2025, are not reflected in this analysis. These developments will undoubtedly shape the future landscape of geoengineering governance and merit careful examination in future scholarship.

K.-U. Schrogl
Institute of Political Science University of Tübingen, Tübingen, Germany
e-mail: kai-uwe.schrogl@gmail.com

C. Giannopapa
Eindhoven University of Technology Eindhoven, Eindhoven, The Netherlands
e-mail: christina.giannopapa@gmail.com

N. Antoni (✉)
Eindhoven University of Technology, Eindhoven, the Netherlands
e-mail: ntorina.antoni@gmail.com

K.-U. Schrogl et al. (eds.), *The Politics of Geoengineering*, Springer Climate,
https://doi.org/10.1007/978-3-032-10095-5_1

governance attention. The chapter concludes by outlining this volume's comprehensive examination of the social, political, and ethical dimensions of climate intervention.

Keywords Geoengineering · Solar Radiation Modification · Carbon Dioxide Removal · Climate Governance · Climate Politics · Risk Assessment · International Relations · Climate Policy

1 What Is Geoengineering and why Will it Matter?

1.1 The Geoengineering Challenge

Geoengineering (hereinafter GE) represents both a potential solution to and a significant risk for addressing the climate crisis (Royal Society, 2009). Similar to Artificial Intelligence (AI), GE encompasses a set of technologies and applications that require careful development and comprehensive regulation (Action Group on Erosion, Technology and Concentration [ETC Group], 2013). However, while regulatory frameworks for AI are gradually emerging, GE remains largely unregulated, positioning it as simultaneously one of the most promising and potentially hazardous approaches to climate intervention.

Contemporary climate policy presents three fundamental pathways for addressing climate change. The first involves drastically reducing carbon-based energy consumption—an approach that, while demanding significant societal changes, offers substantial co-benefits including improved public health and enhanced energy security. The second pathway involves continuing current fossil fuel practices while attempting to manage climate consequences through adaptation measures. The third option centers on active climate intervention through large-scale geoengineering deployment.

Recent international climate negotiations illustrate the trajectory toward geoengineering consideration. The 28th Conference of the Parties (COP28), held in Dubai in 2023, achieved historic agreement to "transition away from fossil fuels"—marking the first time such language appeared in UN climate negotiations, though falling short of mandating complete phase-out (United Nations [UN] Climate Change, 2023). The subsequent decision to hold COP29 in another fossil fuel-producing nation, Azerbaijan, with apparent emphasis on mitigation rather than prevention, suggests an implicit endorsement of geoengineering approaches (Schrogl & Wang, 2024).

This progression toward option three appears increasingly likely when conventional disaster prevention and management investments prove insufficient. However, such deployment risks proceed chaotically due to inadequate preparation across political, legal, regulatory, and societal dimensions.

1.2 Defining Geoengineering Technologies

Geoengineering (GE) constitutes "deliberate large-scale manipulation of the planetary environment to counteract anthropogenic climate change" (Royal Society, 2009:1). This approach fundamentally differs from traditional mitigation strategies focused on reducing greenhouse gas emissions. Instead, GE technologies aim to directly modify Earth's climate systems through two primary methodologies, as assessed extensively in the Intergovernmental Panel on Climate Change (IPCC)'s Fifth Assessment Report (IPCC, 2014). Accordingly, the two main methods in GE are the following:

- Carbon Dioxide Removal (CDR): This method focuses on extracting carbon dioxide from the atmosphere. It can involve natural processes or technological solutions to reduce the amount of atmospheric greenhouse gases.
- Solar Radiation Modification (SRM): This method aims to reflect some of the Sun's energy into space to cool down the Earth's temperature.

CDR encompasses technologies and processes designed to extract carbon dioxide directly from the atmosphere, utilizing both natural and technological approaches (National Academies of Sciences, Engineering, and Medicine [NASEM], 2019). Current research examines multiple dimensions of CDR strategies, with several initiatives already advancing from research phases to implementation. These initiatives span research and development, pilot deployments, and market mechanism development, emphasizing technological advancement, lifecycle assessment, and robust measurement and verification protocols (NASEM, 2019; Frontier Climate, 2023).

SRM technologies aim to reflect portions of solar energy back to space, thereby reducing global temperatures (National Research Council, 2015). Key SRM approaches include cloud brightening through seawater aerosol injection, space-based solar reflectors, and stratospheric aerosol deployment (National Research Council, 2015; IPCC, 2014). The more limited scope of SRM research stems from heightened concerns regarding intervention risks and associated ethical considerations (UNEP, 2023). Current initiatives prioritize research activities, particularly climate modeling, alongside governance frameworks, ethical assessment, and public engagement processes (Carnegie Climate Governance Initiative, 2023).

This analysis employs a Strengths, Weaknesses, Opportunities, and Threats (SWOT) framework to navigate the uncertainties surrounding geoengineering technologies. While SWOT analysis traditionally applies to established technologies, it provides valuable insights when adapted for emerging technologies like GE. Subsequent chapters will present comprehensive SWOT analyses of both CDR and SRM approaches, examining their viability, challenges, and potential contributions to climate change mitigation.

Understanding geoengineering requires recognizing its role as a complement to, rather than substitute for decarbonization and potentially large-scale greenhouse gas removal—not as an alternative to mitigation but as a complementary approach integrated with emissions reduction efforts. While this book primarily emphasizes Carbon Dioxide (CO_2) removal strategies due to CO_2's predominance in anthropogenic emissions and its cumulative long-term climate effects, other potent greenhouse gases also contribute significantly to climate change. Methane (CH_4) deserves particular attention with its global warming potential many times greater than CO_2 over shorter timeframes, with significant sources including permafrost thaw, deep ocean sediments, fossil fuel extraction, and agricultural practices (IPCC, 2021). Other important greenhouse gases include nitrous oxide (N_2O) and fluorinated gases. Our focus on CO_2 reflects the current scientific and policy emphasis, its long atmospheric residence time, and the scale of emissions relative to other greenhouse gases. However, our analysis employs CO_2-equivalent (CO_2e) metrics to implicitly account for all major greenhouse gases.

2 The Climate Action and Geoengineering: Risk Assessment and Governance Evolution

2.1 The Climate Action and Geoengineering Nexus

The relationship between climate action and geoengineering presents a complex risk assessment challenge that requires careful evaluation of competing threats. Current climate trajectories pose substantial risks of irreversible planetary damage, affecting ecosystems, economic systems, and social structures through extreme weather events, biodiversity loss, and socio-economic instability (IPCC 2023; UN Climate Change, 2023). Simultaneously, geoengineering deployment introduces distinct risks stemming from technological uncertainty and limited understanding of long-term environmental impacts (Harvard University, 2024).

This "risk versus risk" framework characterizes the central challenge facing climate policy: balancing the known and escalating consequences of climate inaction against the uncertain and potentially disruptive impacts of geoengineering interventions (World Economic Forum [WEF], 2024). Both CDR and SRM technologies remain in early developmental stages, with potential for unforeseen effects on natural systems. Addressing this dilemma requires comprehensive approaches incorporating robust research protocols, transparent governance mechanisms, and sustained public engagement to evaluate and manage these competing uncertainties.

The IPCC addresses this tension in its Climate Change 2023: Synthesis Report, acknowledging geoengineering's potential contribution to climate mitigation while maintaining strong reservations about substituting these technologies for emissions reduction (IPCC, 2023). While modeling studies demonstrate the conceptual possibility for geoengineering to help limit warming below 1.5 °C, the IPCC emphasizes the availability of proven emission reduction strategies, including renewable energy deployment, energy efficiency improvements, and consumption pattern modifications.

2.2 International Governance Responses

International governance efforts reflect the complexity of geoengineering oversight. The United Nations Environment Assembly (UNEA) has approached geoengineering with considerable caution, highlighting regulatory complexities. Switzerland's 2024 proposal to establish a study group on SRM was withdrawn after failing to secure sufficient support, revealing deep international divisions over governance approaches (McLaren & Corry, 2024). The February 2024 UNEA-6 meeting in Nairobi demonstrated these divisions, with African countries led by Djibouti advocating for "global governance mechanisms for non-use", while the US and Saudi Arabia opposed the proposal and pushed for limited oversight, with the EU, Pacific Island states, Colombia, and Mexico supporting the African Group's position (Lo, 2024; McLaren & Corry, 2024).

Current regulatory efforts face significant obstacles, particularly from high-emission and oil-producing nations (Heinrich Böll Stiftung, 2019). However, limited progress exists through de facto moratoriums on marine geoengineering techniques such as ocean fertilization under the London Convention and London Protocol (Geoengineering Monitor, 2021). The United Nations Environment Programme's (UNEP) 2023 report "One Atmosphere: An Independent Expert Review on Solar Radiation Modification Research and Deployment" concluded that large-scale solar radiation modification is "not necessary, viable, prudent or sufficiently safe" (UNEP, 2023).

3 From Early Visions to Politics: The Past and Present of Conceptualising Geoengineering—A Model Case for Multi- and Interdisciplinary Approaches

3.1 Conceptual Foundations

Geoengineering concepts have evolved through diverse intellectual contributions spanning scientific, technical, and literary domains. Hermann Oberth's 1923 work "Die Rakete zu den Planetenräumen" introduced space-based solar reflectors as climate intervention tools, proposing mirror systems 100–300 km in diameter arranged in adjustable orbital networks. Nobel laureate Paul J. Crutzen significantly advanced geoengineering discourse through his atmospheric chemistry research and introduction of the "Anthropocene" concept in 2000, characterizing the epoch of significant human environmental impact.

Crutzen's 2006 paper "Albedo Enhancement by Stratospheric Sulfur Injections" marked a pivotal moment in legitimizing geoengineering research, advocating serious investigation of stratospheric aerosol injection as climate intervention. This work transformed previously fringe scientific concepts into mainstream research considerations. Contemporary literary works, including Kim Stanley Robinson's "The Ministry for the Future", explore geoengineering's societal, ethical, and environmental implications, contributing to broader public discourse on technological climate solutions.

3.2 Scientific Assessment and Research Infrastructure

Recent scientific assessments have substantially advanced geoengineering understanding through comprehensive interdisciplinary evaluation. The IPCC's Fifth Assessment Report (IPCC, 2014) provided systematic analysis of geoengineering feasibility, risks, and potential benefits across multiple dimensions. NASEM contributed specialized assessments on solar radiation modification (National Research Council, 2015) and carbon dioxide removal (2019), focusing on specific technological approaches and implementation considerations (NASEM, 2019).

The Royal Society's 2009 report "Geoengineering the Climate: Science, Governance, and Uncertainty" established foundational understanding of geoengineering science, practical applications, and ethical considerations (Royal Society, 2009). The Carnegie Climate Governance Initiative has contributed to developing governance frameworks, ethical considerations, and societal impacts (Carnegie Climate Governance Initiative, 2023).

3.3 Research Institutes and Programs

Research institutions have established specialized programs addressing both technical and governance dimensions of geoengineering. The Oxford Geoengineering Programme has contributed technical analysis and governance frameworks, influencing policy development and public dialogue. Harvard University's Solar Geoengineering Research Program investigated marine cloud brightening, cirrus cloud modification, space-based interventions, and stratospheric aerosol deployment. However, Harvard announced in March 2024 the discontinuation of the Stratospheric Controlled Perturbation Experiment (SCoPEx) following sustained international criticism and repeated delays (Temple, 2024a).

This institutional shift reflects broader changes in the geoengineering research landscape, exemplified by David Keith's 2023 transition to the University of Chicago to establish a new Climate Systems Engineering initiative (University of Chicago News, 2023). The GeoEngineering Centre at Queen's University and Royal Military College of Canada continues contributing to technical research, while the Oxford Martin School's 2010 "Oxford Principles for Geoengineering Governance" provided foundational ethical guidelines for research conduct and implementation (Rayner et al., 2013). The International Risk Governance Center (IRGC) at the Swiss Federal

Institute of Technology (EPFL) published a report in 2020, which offers policy recommendations for an international regulatory framework for GE.

3.4 International Governance Initiatives

Governance initiatives demonstrate increasing emphasis on multistakeholder and interdisciplinary approaches to geoengineering oversight. The Solar Radiation Management Governance Initiative (SRMGI), launched in March 2010 through collaboration between the Royal Society, TWAS (The World Academy of Sciences), and Environmental Defense Fund, promotes international dialogue and supports developing country capacity building (TWAS, n.d). Through the DECIMALS fund, SRMGI has supported Global South modeling research analyzing regional solar radiation management impacts (TWAS, 2018).

The Alliance for Just Deliberation on Solar Geoengineering, established in April 2023, represents innovative governance approaches emphasizing procedural justice and democratic participation (SGDeliberation.org, 2023; Kleinman Center for Energy Policy, 2024). This Washington D.C.-based organization explicitly avoids advocacy positions, focusing instead on enabling inclusive deliberation by empowering climate-vulnerable communities and civil society organizations in governance framework development.

The Planetary Sunshade Foundation represents recent developments in space-based solar radiation management research. In 2024, the Foundation received £400 k funding from the UK's Advanced Research and Invention Agency (ARIA) for collaborative work with Cornell University, National Center for Atmospheric Research, University of Nottingham, Redwire Space, and Jet Propulsion Laboratory (SRM360, 2024). This collaboration advances space-based reflector concepts through rigorous engineering analysis and climate modeling of atmospheric dynamics, chemistry, and oceanic feedback systems (Planetary Sunshade Foundation, 2024).

3.5 Civil Society and Policy Responses

Civil society engagement reflects increasing polarization around geoengineering research and deployment. Over 600 scientists from more than 60 countries have endorsed a Solar Geoengineering Non-Use Agreement, calling for restrictions on public funding, patent protection, experimental activities, and international support for solar geoengineering research (Open Letter on Solar Geoengineering Non-Use Agreement, 2022). This scientific opposition emphasizes concerns about long-term impacts and ethical implications of large-scale climate interventions (Preston, 2013).

The Climate Overshoot Commission, comprising sixteen international experts and former global leaders, has evaluated risks associated with significant climate overshoot scenarios and their implications for geoengineering governance. The Commission's September 2023 report recommended prioritizing emissions reduction and fossil fuel phase-out while calling for enhanced international discussions and scientific research on SRM technologies (Climate Overshoot Commission, 2023). Key recommendations include establishing governance frameworks for SRM research while maintaining moratoriums on deployment and large-scale outdoor experiments posing transboundary risks.

Organizations worldwide have adopted sharply divergent positions on geoengineering governance. The Center for International Environmental Law (CIEL) advocates for an international Solar Geoengineering Non-Use Agreement, calling on governments to ban all outdoor experiments, prohibit public funding for solar geoengineering research, prevent technology development and patenting, and uphold the de facto moratorium established under the Convention on Biological Diversity (CIEL, 2023). In contrast, scientific bodies such as the U.S. National Academies of Sciences, Engineering, and Medicine have called for expanded research programs under governance frameworks, arguing that better understanding of solar radiation modification risks and benefits is essential for informed decision-making (National Academies, 2021). The Institute for Advanced Sustainability Studies (IASS) has facilitated international dialogue through its Climate Engineering Conference series, providing forums for diverse stakeholder engagement on both carbon dioxide removal and solar radiation modification research developments (IASS, 2017). These divergent organizational perspectives underscore the polarized nature of the geoengineering governance debate, with growing civil society opposition occurring alongside continued research funding expansion and increased scientific and policy attention to potential climate intervention technologies.

These diverse organizational perspectives underscore the importance of collaborative, global approaches to addressing scientific, ethical, and governance challenges posed by geoengineering technologies. However, the landscape remains increasingly polarized, with growing civil society opposition occurring alongside continued research funding expansion and private sector interest.

4 The Status of the Policy Debate on Geoengineering

4.1 In the EU

The European Union has emerged as a decisive voice in geoengineering governance, adopting increasingly restrictive positions on solar radiation modification technologies. In June 2023, the European Commission and EU External Action Service issued a joint communication on the climate-security nexus, explicitly stating that the EU does not consider geoengineering a viable climate solution absent clear scientific understanding (European Commission, 2023). The Commission stated that SRM technologies introduce new risks to people and ecosystems with poorly understood impacts and unintended consequences, while calling for global governance framework discussions emphasizing transparency, developing country inclusion, and adherence to precautionary principles. The Commission's position was further clarified in its August 2023 scoping paper, which stated that "SRM deployment represents an unacceptable risk for humans and the environment" in the current state of development (European Commission, Group of Chief Scientific Advisors, 2023).

This position was reinforced in December 2024 when the EU's Group of Chief Scientific Advisors released comprehensive SRM recommendations, advocating against deployment of SRM technologies while supporting continued research under strict governance frameworks with periodic reassessment cycles of five to ten years (European Commission, Group of Chief Scientific Advisors, 2024). The advisors recommended Europe-wide moratoriums on SRM deployment while maintaining prioritization of emissions reduction and climate adaptation strategies.

Parallel to these restrictions, the EU has advanced comprehensive carbon management strategies. The February 2024 EU industrial carbon management strategy establishes ambitious targets of capturing approximately 280 million tonnes of CO_2 by 2040 and 450 million tonnes by 2050, with significant contributions from biogenic sources and atmospheric capture (European Commission, 2024). Industry stakeholders have expressed capacity to capture 80 million tonnes annually by 2030 given appropriate investment conditions (Zero Emissions Platform, 2024).

The EU strategy encompasses three primary pathways: Carbon Capture and Storage (CCS) for permanent geological storage of captured emissions; Atmospheric Carbon Removal for permanent storage of biogenic or atmospheric CO_2; and Carbon Capture and Utilisation (CCU) for incorporating captured CO_2 into synthetic products, chemicals, and fuels with strategic emphasis on biogenic or atmospheric sources for enhanced climate benefits.

4.1.1 Multilateral European Initiatives

Building on these EU-wide policy frameworks, European implementation of carbon management infrastructure has advanced through multiple mechanisms. In November 2023, Denmark, Germany, Sweden, France and the Netherlands signed the Aalborg Declaration recognising carbon capture, utilisation and storage as a critical tool for the green transition and achieving net-zero emissions by 2050, facilitating collaborative project development and calling for cost-effective cross-border CO_2 transport infrastructure and improved regulatory predictability for permanent negative emissions (Danish Ministry of Climate, Energy and Utilities, 2023). This initiative facilitates collaborative project development and cost-effective CO_2 transportation systems between countries while calling for improved regulatory predictability for permanent negative emissions.

Simultaneously, several EU countries have withdrawn from the 1998 Energy Charter Treaty due to concerns about impediments to climate action. Denmark,

France, Germany, Luxembourg, Poland, Spain, and the Netherlands have opted out, concerned about treaty provisions allowing energy companies to challenge government climate policies through legal action (European Parliament, 2023).

4.1.2 National Developments

While multilateral European initiatives demonstrate regional coordination, individual member states have developed distinct national approaches that reflect both shared EU principles and specific national circumstances.

The United Kingdom's £56.8 million Advanced Research and Invention Agency (ARIA) investment in solar engineering research, announced in 2024, represents the largest single source of such funding globally and marks a significant escalation in research scope (SRM360, 2024). Unlike earlier research programs that focused exclusively on modeling, the ARIA programme explicitly includes funding for small-scale outdoor experiments across multiple countries, including Arctic sea ice thickening, marine cloud brightening experiments in Australia and the UK, stratospheric aerosol injection sampling in the US or UK, and natural materials testing for stratospheric aerosol injection. This approach contrasts sharply with China's earlier government-funded research program (approximately $three million federal funding across three institutions from 2015–2019), which explicitly avoided outdoor experiments and focused instead on climate modeling, impact assessment, and governance issues (Temple, 2017). The divergent approaches highlight evolving international positions on the appropriate scope and methods of geoengineering research. UK carbon management infrastructure has advanced through regulatory mechanisms, with the North Sea Transition Authority (NSTA) awarding 21 carbon storage licenses to 14 companies in September 2023, covering approximately 12,000 square kilometers of depleted oil and gas reservoirs and saline aquifers with potential storage capacity of 30 million tonnes of CO_2 per year by 2030—approximately 10% of UK annual emissions (NSTA, 2023). In January 2024, the UK Government approved Drax Power Station's Development Consent Order for converting two biomass units to bioenergy with carbon capture and storage (BECCS), enabling potential removal of approximately 8 million tonnes of CO_2 per year when both units become fully operational (UK Government, 2024; Drax Group, 2024).

Germany's December 2023 Climate Foreign Policy Strategy affirmed the government's position that solar radiation modification technologies do not constitute viable climate policy options given existing uncertainties, implications, and risks (Auswärtiges Amt, 2023). This position has been consistently maintained by the German government, which continues its commitment to the 2010 international decision establishing a broad moratorium on geoengineering under the Convention on Biological Diversity. However, German think tanks have only gradually begun to consider geoengineering in their research, particularly given growing private sector interests in the field (Schrogl, 2022). The German Environment Agency advocates for a moratorium on geoengineering activities until a sufficient scientific foundation is established to justify their implementation (German Environment Agency, 2023).

In November 2023, Switzerland's Federal Council laid the groundwork for the export of CO2 for sub-seabed storage by deciding to ratify the 2009 amendment to Article 6 of the London Protocol (Swiss Federal Council, 2023). Switzerland's proposal at the February 2024 UN Evironment Assembly (UNEA-6) to establish an expert panel on solar radiation modification was withdrawn following strong opposition from the African Group, Pacific Island states, and Latin American countries (McLaren & Corry, 2025).

4.2 In the United States

In the United States, government engagement with GE research has expanded significantly. The National Oceanic and Atmospheric Administration (NOAA) has received over $22 million over three years for research programs that include efforts to assess "climate interventions" through the Earth's Radiation Budget Initiative (Temple, 2022). This funding represents a substantial increase from earlier estimates and includes extensive balloon-based monitoring networks across five locations globally. In response to a congressional mandate, the White House Office of Science and Technology Policy released a five-year research plan in June 2023 (The White House, 2023).

The White House report, developed in coordination with NOAA and other federal agencies including NASA and the Department of Energy, identifies knowledge gaps and potential research areas while emphasizing the importance of understanding risks and governance challenges rather than supporting deployment, marking the first comprehensive U.S. government position on solar geoengineering. The report also addresses the importance of research capabilities to detect potential SRM deployments and geopolitical considerations associated with these technologies (The White House, 2023).

4.3 In China

In China, the government has shown interest in GE as a potential tool for climate change mitigation and adaptation (Sharma, 2023). The State Council of the People's Republic of China (PRC) announced an ambitious plan to develop a "weather modification system" by 2025, covering 5.5 million square kilometers of its land area through cloud seeding, a technique developed to increase agricultural productivity and prevent natural disasters such as droughts and floods (The State Council of PRC, 2020).

China has built one of the world's largest geoengineering research programs, involving around 15 faculty members and 40 students across three institutions: Beijing Normal University, Zhejiang University, and the Chinese Academy of Sciences (Temple, 2017). The program completed its major government-funded phase in 2017, having emphasized international cooperation and multilateral governance approaches rather than unilateral development (Temple, 2017).

However, China is not a full signatory to the Convention on the Prohibition of Military or Any Other Hostile Use of Environmental Modification Techniques (ENMOD), though it agreed in 2005 that the Convention would apply to Hong Kong and Macao Special Administrative Regions (International Committee of the Red Cross, 2005). Researchers involved in China's solar geoengineering programme have emphasized that current work is limited to computer models and laboratory studies and that China "will not take action before an international consensus has formed" (Temple, 2017). The programme's public materials similarly stress that it does not conduct any "experiments on our real planet" but instead explores the effectiveness and uncertainties of SRM through computer modelling (Jayaram, 2019). Concerns have been raised about the potential transboundary impacts of these techniques, particularly cloud seeding operations that may affect weather patterns in neighboring countries such as India, Nepal, Laos, Myanmar, and other Mekong-Brahmaputra basin states (The State Council of PRC, 2020; Jayaram, 2019).

4.4 Global Policy Responses

While major powers like the EU, United States, and China have begun developing distinct approaches to geoengineering governance, the broader international community has also started responding to geoengineering developments, often in reaction to specific triggering events.

A watershed moment occurred on January 13, 2023 when Mexico became the first country to explicitly announce measures to ban outdoor solar geoengineering experiments following rogue deployments by startup company Make Sunsets, which had been releasing sulfur particles into the atmosphere without government permission (Rodriguez & Lo, 2023; Temple, 2022). This unprecedented national prohibition demonstrated how private sector experimentation could force rapid policy responses and highlighted the governance vacuum around commercial geoengineering activities.

Other countries have also made significant contributions and are growing in number and influence. These nations play an increasingly important role in shaping the political agenda and institutional frameworks surrounding geoengineering. The African Group has emerged as a coordinated voice in international forums, particularly through their position at UNEA-6 in February 2024 calling for comprehensive risk assessment and precautionary approaches (Lo, 2024). This represents more sophisticated Global South engagement than previously seen, including specific technical expertise and legal argumentation rather than simple opposition.

This growing engagement from developing countries has been facilitated through targeted capacity-building efforts and regional dialogue initiatives. Since ealry 2010's, regional workshops have been held in countries like Bangladesh, Brazil, China, Ethiopia, India, Jamaica, Kenya, Thailand, New Zealand (for Pacific states),

Pakistan, and the Philippines, bringing together local climate scientists, journalists, policymakers and civil-society representatives to learn about and debate solar geoengineering (Rahman et al., 2018). These workshops provide platforms for scientists, policymakers, and stakeholders to explore the potential benefits, risks, and ethical implications of SRM. Small Island Developing States' (SIDS) have increasingly emphasized the need for meaningful participation in SRM governance. The Americas Conference on Solar Radiation Modification: Science, Governance and Implications for the Region, held in Kingston, Jamaica in August 2022, highlighted that SIDS should be actively involved in shaping any SRM research and governance frameworks, rather than being treated as passive consultees (Scobie, 2022).

Two initiatives that exemplify these efforts are the Solar Radiation Management Governance Initiative (SRMGI) and the Degrees Initiative. SRMGI—launched by the Royal Society, Environmental Defense Fund and TWAS—has organised workshops in 14 countries and funded research teams in Argentina, Bangladesh, Benin, Indonesia, Iran, Côte d'Ivoire, Jamaica and South Africa, enabling Global South researchers to model how SRM might affect regional climate risks (NASEM, 2021).

As outlined, the conceptualisation and development of geoengineering have been influenced by a range of factors, from early visions to modern scientific research, policy debates, and portrayals in literature. The period from 2023–2025 has seen unprecedented acceleration in both governmental engagement and civil society opposition, creating an increasingly polarized landscape that will continue to shape the future of geoengineering governance. This evolving political landscape reveals that geoengineering governance cannot be understood through policy analysis alone. The complex interplay of environmental risks, international politics, economic interests, legal frameworks, and social concerns requires a multidimensional analytical approach that examines the key factors shaping geoengineering development and deployment.

5 Key Dimensions of GE Analysis

To effectively navigate the complexities revealed in the policy landscape above, it is essential to examine the key dimensions that shape geoengineering development and deployment. Understanding geoengineering requires systematic examination across multiple interconnected dimensions, each presenting distinct challenges for potential climate interventions.

The *environmental dimension* focuses on how geoengineering affects planetary systems, including how techniques that adjust incoming solar radiation or atmospheric CO_2 concentrations may disrupt environmental processes. Analyses and reviews highlight substantial risks and trade-offs across different approaches, from land- and water-intensive carbon-dioxide removal to the climatic and ecological side-effects of solar radiation modification (Dunne, 2018; Kaufman, 2019).

For Solar Radiation Modification, this involves investigating optimal particle types and distribution methods to minimize weather pattern disruption. Environmental assessment requires understanding complex feedback mechanisms, potential tipping points, and long-term ecological consequences that may emerge from large-scale climate interventions.

The *international political dimension* addresses challenges concerning global cooperation and governance in GE. The February 2024 failure of Switzerland's UN resolution and ongoing international forum deadlocks demonstrate deep divisions between countries regarding governance approaches for geoengineering research and potential deployment.

These divisions reflect broader geopolitical tensions, with some countries viewing geoengineering as potential climate colonialism while others consider it necessary climate risk management. The concentration of research capacity in Global North institutions, combined with philanthropic funding dominated by wealthy individuals, creates power structures viewed with suspicion by many developing countries.

Governance requires identification of suitable frameworks, drawing lessons from other global resource governance domains (Talberg et al., 2018). Geoengineering governance research focuses on developing international treaties and agreements ensuring fair participation principles and compliance from all nations (Biermann et al., 2022). Effective governance requires diplomatic processes and proactive mechanisms ensuring global consensus and compliance (McEvoy et al., 2024).

Beyond the political complexities of international coordination, the *economic dimension* of GE involves various technologies with differing costs and uncertainties (Moreno-Cruz & Smulders, 2017). Carbon Dioxide Removal costs vary significantly by method, with Bioenergy with Carbon Capture and Storage (BECCS) and Direct Air Capture (DAC) differing in cost, scalability, and atmospheric CO_2 removal effectiveness (NASEM, 2019; IEA, 2022). Solar Radiation Modification costs include technology expenses and potential unintended effects such as altered rainfall patterns. Limited SRM research complicates comprehensive cost-benefit analyses, creating uncertainty about economic implications of large-scale deployment.

Oil company involvement in Carbon Capture and Storage initiatives raises concerns about potential conflicts of interest, where climate problem contributors might benefit disproportionately from solutions while potentially using these technologies to justify continued fossil fuel production rather than genuine decarbonization (Chalmin, 2021; Drugmand, 2023).

Billionaire participation in geoengineering research funding presents both opportunities and risks, potentially accelerating progress while raising concerns about research direction influence, personal gain prioritization over public benefit, and reduced urgency for greenhouse gas emission reductions through viewing geoengineering as potential quick fixes (De la Garza, 2023).

The *legal and regulatory dimension* concerns the laws and regulations governing GE deployment. The economic considerations outlined above intersect significantly with legal and regulatory dimensions that concern laws and regulations governing geoengineering deployment. Mexico's 2023 ban on solar geoengineering experiments represents the first national prohibition and demonstrates how governance gaps can lead to reactive policy responses (Rodriquez & Lo, 2023; Lo, 2024). The absence of comprehensive international legal frameworks means that most governance occurs through voluntary guidelines and research ethics protocols, creating significant regulatory uncertainty for both researchers and the public.

Absent specific geoengineering legal frameworks, governance development can learn from other emerging technology challenges, such as biotechnology, tailoring regulations that encourage innovation while ensuring responsible deployment (OECD, 2024; Hands Off Mother Earth Campaign, 2022).

While legal and regulatory frameworks provide formal governance structures, *societal and public dimension* addresses public perception and engagement that ultimately determine the legitimacy and feasibility of geoengineering approaches. Growing civil society organization opposition, including the Hands Off Mother Earth Alliance and various global resistance networks, demonstrates significant public resistance, with over 600 scientists from more than 60 countries endorsing calls for a Solar Geoengineering Non-Use Agreement (Geoengineering Monitor, 2024; Open Letter on Solar Geoengineering Non-Use Agreement, 2022).

Inclusive public engagement challenges have led to new institutional approaches, such as the Alliance for Just Deliberation on Solar Geoengineering, established to build governance capacity and elevate climate-vulnerable region voices without taking deployment advocacy positions. Effective public engagement requires making geoengineering concepts comprehensible to broader audiences, fostering informed dialogue through clear, transparent communication shaping public understanding about technology deployment (Kahan et al., 2015; Scheer & Renn, 2014). This involves commitment to clear dialogue and development of communication techniques resonating with public concerns and aspirations.

Underlying all the dimensions are fundamental questions of justice and equity that remain critically underdeveloped in current governance approaches. The practical impossibility of implementing Free, Prior, and Informed Consent for planetary-scale interventions represents a fundamental challenge to indigenous rights (Buck et al., 2014; Whyte, 2018). Climate colonialism dynamics risk perpetuating global inequalities through new mechanisms of environmental control, while the termination problem—risks created by sudden cessation of geoengineering programs—poses serious intergenerational justice challenges that current governance frameworks inadequately address.

6 Theories and Methodologies for the Politics of GE into the 2030s

To understand the politics of GE, this book adopts a multidisciplinary approach integrating insights from political science, economic science, and innovation science perspectives, while incorporating emerging conceptual frameworks such as feminist foreign policy theory to illuminate the evolution and implementation of geoengineering governance.

Political science provides essential insights into the governance structures and policy-making processes that shape geoengineering development. By applying established international relations theories—including realism, liberalism, and constructivism—we can better understand the complex interactions, institutional frameworks, and regulatory mechanisms necessary for effective geoengineering governance. From a realist perspective, state behavior in geoengineering reflects traditional power competition dynamics, where countries pursue technological capabilities to enhance national interests and gain strategic advantages (Mearsheimer, 2003). The UK's £56.8 million ARIA 'Exploring Climate Cooling' investment (SRM360, 2024) and China's Ministry of Science and Technology–funded ~17 million yuan solar geoengineering research program launched in 2015 (Temple, 2017) exemplify this competitive dynamic, where states pursued geoengineering capabilities to enhance climate security while potentially establishing technological superiority. Liberal institutionalism offers a contrasting framework, emphasizing the potential for international cooperation and rule-based governance in addressing global challenges (Keohane & Nye, 1973). However, the failure of multilateral initiatives such as the Swiss UN resolution demonstrates the limitations of liberal institutionalist approaches when fundamental national interests and value systems diverge significantly. As constructivist scholars emphasize perceptions of risk and legitimacy shape how technologies are governed (Wendt, 1999). In solar geoengineering, critical work shows how framings by powerful actors can normalise or contest the technology, with significant implications for policy outcomes (Stephens & Surprise, 2020). A crucial question in the governance literature concerns whether geoengineering will emerge as a distinct international regime with specialized rules and processes, following International Regime Theory, or whether it will be integrated within existing frameworks addressing environmental protection, security, or global commons governance (Möller, 2020). Political science analysis must identify and evaluate conceptual and organizational approaches that yield optimal governance outcomes across these different institutional arrangements.

Several **contemporary political science and philosophical approaches** offer innovative analytical perspectives for understanding geoengineering governance. Achille Mbembe's concept of the "politics of enmity" provides valuable insight into potential governance conflicts, suggesting that modern liberal democracies are characterized by pervasive desires for separation and control that could manifest in geoengineering as divisions between those possessing technological control and those subject to environmental impacts (Mbembe, 2019). Chinese international political philosopher Zhao Tingyang's concept of "tianxia" (all under heaven) offers an alternative governance horizon emphasizing global unity and inclusive decision-making rather than competitive state interests (Babones, 2020, as cited in Zhao, 2021). This approach suggests possibilities for more collaborative geoengineering governance that transcends traditional sovereignty-based international relations. Feminist Foreign Policy theory introduces additional analytical dimensions by promoting gender equality, women's rights, peace, environmental sustainability, and human rights while challenging traditional power systems (Björkdahl & Selimovic,

2019). Applied to geoengineering governance, these perspectives emphasize inclusive decision-making processes and attention to differential impacts across gender, class, and regional lines.

Economic analysis provides essential tools for understanding geoengineering's value creation, cost structures, and market dynamics. Different economic frameworks offer distinct insights into how geoengineering technologies might develop and be deployed. A neoclassical perspective would suggest that increased demand for geoengineering technologies could drive cost changes through standard supply-demand dynamics, yet the current philanthropic funding model complicates this picture. Funding is dominated by wealthy individuals and specialised foundations-such as Bill Gates and the Quadrature Climate Foundation's pledge of $40 million investment-which demonstrates how concentrated wealth can shape research priorities outside normal market mechanisms, creating unique political-economic dynamics (Temple, 2024b). Keynesian approaches emphasize the role of government intervention to maintain market stability and address market failures (Krugman, 2009). In geoengineering contexts, this framework supports public sector involvement in establishing ethical guidelines, ensuring alignment with sustainable development goals, and managing the public goods characteristics of climate interventions. Behavioral economics examines how psychological factors and cognitive biases affect economic decision-making processes (Kahneman & Tversky, 2013). Applied to geoengineering, this perspective reveals how public-safety concerns, risk perceptions, and trust or mistrust in institutions could significantly impact financial support and technology adoption, regardless of potential climate benefits.

Political economy perspectives examine how different interest groups and social factors shape policy development and resource allocation. The oil industry's substantial involvement in Carbon Capture and Storage (CCS) and other Negative Emission Technologies (NETs) illustrates how such approaches can lock in fossil infrastructures, help reduce stranded-asset risks and reduce the pressure for rapid emissions reductions, a dynamic often described as "mitigation deterrence" (Scherrer et al., 2023; Carton, 2019; Markusson et al., 2018). Mariana Mazzucato's mission-oriented innovation approach suggests fundamental rethinking of government-business relationships in emerging technologies, ensuring that public investments create and retain public value, supporting equitable and sustainable outcomes rather than merely privatized benefits (Mazzucato, 2021). Applied to geoengineering, this framework emphasizes the need for governance structures that ensure public value from geoengineering research and development is broadly shared and aligned with societal goals, rather than captured narrowly by incumbent fossil-fuel interests.

Innovation science provides frameworks for understanding the dynamics of technological development and diffusion in geoengineering. Multiple theoretical approaches illuminate different aspects of innovation processes and their social implications (Fagerberg et al., 2005). Rogers' diffusion-of-innovations theory explains how geoengineering technologies might spread through society, depending on factors such as perceived usefulness, complexity, and observability (Rogers et al., 2014).

However, this traditional framework may inadequately account for the significant socio-political and economic barriers that can impede adoption of controversial technologies like geoengineering (Boettcher & Schäfer, 2017). By analogy with disruptive innovation theory (Christensen, 1997), geoengineering technologies could fundamentally transform approaches to climate-change mitigation and adaptation. Yet, this disruptive potential, as patents indicate, generates significant ethical and governance challenges, from uncertainty about unintended climatic consequences to concerns that emerging patents, trials and investments may exacerbate global inequalities, particularly for the Global South (NASEM, 2021; Ramos & Santos, 2025). Open innovation theory emphasizes utilizing ideas and capabilities from both internal and external organizational boundaries (Chesbrough, 2003). For geoengineering, this lens suggests that collaboration among universities, governments, industry, and civil society can develop more effective and socially acceptable technologies, but only if supported by transparent and inclusive governance frameworks that are often lacking in current institutional arrangements (Ramos & Santos, 2025).

Philosophical analysis addresses the fundamental ethical and moral challenges that geoengineering presents, providing essential foundations for governance framework development. Existentialist approaches examine how geoengineering might fundamentally alter human identity and our relationship with planetary systems, raising questions about authenticity and human agency in environmental management (Sartre, 2018; Hamilton, 2013). This perspective highlights the profound ontological implications of deliberate climate intervention. Utilitarian frameworks attempt to systematically weigh potential benefits, such as reducing global warming impacts, against risks including unpredictable weather changes and ecosystem disruption (Mill, 1863; Singer, 2011). However, conducting meaningful utilitarian calculations remains challenging given the long-term temporal scales and global spatial scales that make impacts difficult to predict and quantify accurately (Callies, 2023). Deontological approaches question whether humanity possesses the moral right to deliberately manipulate planetary systems regardless of potential climate benefits, emphasizing moral obligations that exist independent of consequential calculations (Kant, 1785; Moore & Alexander, 2007). This framework raises fundamental questions about human stewardship and planetary boundaries.

Drawing parallels with **AI governance** offers valuable insights for geoengineering regulation, given both technologies' complexity and far-reaching implications requiring effective regulatory frameworks. Algorithmic accountability theory emphasizes transparency, explainability, and responsibility in AI systems (AI Now Institute, 2023). Applied to geoengineering, this framework implies essential transparency requirements enabling all stakeholders to understand how technologies function and evaluate their effectiveness and safety. AI ethics frameworks focus on fairness, privacy, and human rights protection. When applied to geoengineering, these principles emphasize ensuring technologies serve the common good rather than narrow interests (Floridi et al., 2018). This involves preventing disproportionate benefits for certain groups while harming others and respecting the rights and dignity of all affected individuals.

The Stockholm International Peace Research Institute's 2023 study on AI for climate security, examining the intersection of artificial intelligence and climate governance, suggests the need for equivalent analysis regarding AI applications in geoengineering (Kim & Boulanin, 2023). Such analysis would explore both regulatory lessons from AI governance and responsible AI applications in geoengineering systems. The European Union's AI Act (Regulation EU 2024/1689), which entered into force in August 2024, provides a potentially useful model for developing risk-based regulatory frameworks for geoengineering (European Parliament and Council of the European Union, 2024). However, critical assessment of these parallels remains essential. While AI and geoengineering share certain governance challenges—particularly managing emerging technologies with potentially global impacts—geoengineering poses unique intergenerational risks and ethical dilemmas due to its potential for direct global climate system alteration, making governance stakes and requirements, in important respects, different from AI regulation (ETC Group, 2013).

Recent developments in geoengineering governance reveal significant gaps in existing theoretical frameworks that require analytical attention and conceptual development. The Mexico ban on solar geoengineering experiments demonstrates how unilateral private sector actions can force reactive governance responses, suggesting the need for anticipatory governance frameworks that address private sector experimentation before it occurs. The UK's major funding commitment occurring simultaneously with EU scientific advisory bodies calling for a moratorium on solar geoengineering and highly restrictive conditions on SRM research (SRM360, 2024) illustrates how theoretical frameworks must account for divergent national approaches even within traditionally allied countries. The emergence of organized opposition through networks such as the Hands Off Mother Earth Alliance and the Alliance for Just Deliberation on Solar Geoengineering's procedural justice approach demonstrates how civil society organizations are developing sophisticated governance alternatives outside traditional state-centered frameworks. These developments suggest the need for theoretical frameworks that better account for non-state actor influence and alternative governance models.

In conclusion, a comprehensive multidisciplinary approach combining political science, economic science, innovation science, and philosophical frameworks provides essential analytical tools for understanding geoengineering politics into the 2030s. The period from 2023–2025 has demonstrated that existing theoretical frameworks must evolve significantly to address the rapidly changing governance landscape, including unprecedented private sector involvement, organized civil society opposition, and divergent governmental responses. Future theoretical development must better account for non-state actor influence, indigenous sovereignty challenges, and the unique governance requirements of planetary-scale interventions that exceed traditional international relations theory scope. Additionally, frameworks must address the ethical implications of technologies that could fundamentally alter human-environment relationships while ensuring that geoengineering implementation—if it occurs—proceeds through ethical, equitable, and sustainable governance processes.

7 Key Topics Covered

The topics selected in this book draw on the above reflections about the importance of GE and the growing interest it has sparked in research. Our contributors, drawing from existing literature, focus on these dimensions and propose new directions to broaden the scope of investigation. In order to provide the reader with the opportunity to compare the topics contained in the thematic chapters in this book, the contributions follow six steps. The first step assesses the relevance and importance of the topic. The second step provides a definition and identifies key problems and conflicts related to the specific topic. The third step presents the actors and types of institutions involved, including their positions. The fourth step identifies policy and regulatory approaches with possible solutions by examining existing measures and considering additional options that should be introduced to effectively address the respective issue. The fifth step explores policy and diplomatic mechanisms to reach solutions, proposing new and innovative approaches. Finally, the sixth step contributes to the politics of GE by analyzing the particular topic and its impact on political science, international relations, and international law, among others. The chapters slightly vary in the application of this structure, while overall comparability is assured.

Valeria Pinna, *Laure-Marine Vioujard*, *Leonora Burnik*, and editors provide in Chapters 2 and 3 a comprehensive SWOT analysis (Strengths, Weaknesses, Opportunities, Threats) for CDR and SRM respectively. The SWOT analysis for CDR examines its unique strengths, like carbon sequestration potential, and weaknesses, including technological limitations and possible environmental risks. Opportunities for integrating CDR into climate change strategies are discussed, along with potential threats such as economic feasibility and public perception. By juxtaposing these analyses, the chapters provide a nuanced understanding of the multifaceted landscapes of SRM and CDR, offering valuable insights to navigate responsible decision-making in the context of climate intervention strategies. Similarly, for SRM, strengths such as rapid global warming mitigation are weighed against weaknesses like uncertainties in long-term impacts and ethical concerns. Opportunities for SRM deployment, such as its potential role in climate change mitigation, are explored, alongside potential threats like geopolitical tensions and unintended environmental consequences.

Joshua Horton explores the policy implications of CDR options and SRM in his chapter "Governing Carbon Dioxide Removal and Solar Radiation Modification". The author identifies the governance challenges for large-scale, permanent CDR, which include avoiding the moral hazard of undermining emissions reductions, managing side effects, ensuring permanence, and incentivizing innovation and upscaling. The role of international and national/regional bodies in the governance of CDR is also discussed. The chapter further delves into the environmental and political risks associated with SRM and emphasizes the need for international cooperation for their effective management. The chapter highlights the absence of contemporary global governance on SRM and discusses the potential political divide within the Global North regarding SRM. In conclusion, Horton suggests that the future of SRM is likely to be influenced by the increasingly competitive relationship between the US and China. This comprehensive analysis provides valuable insights into the complex dynamics of climate change mitigation strategies.

In the chapter titled "Asia-Pacific Perspectives on Solar Geoengineering", *Masahiro Sugiyama* investigates research, governance activities, and public perceptions related to solar GE in the Asia-Pacific region. He notes that despite significant progress in research and governance, these activities are predominantly concentrated in high-income countries, creating a bias toward the Global North. Emerging literature reveals geographical differences in public perceptions, cultural traditions, and attitudes toward large-scale engineering. Some studies indicate that the public in the Global South is more receptive to the idea of solar GE than those in the Global North. However, all regions exhibit caution about research and deployment and favor global governance. The Asia-Pacific regions have initiated various regional intervention research projects, such as "Typhoon Shot" in Japan and "Sky River" in China. To incorporate diverse perspectives, the chapter advocates for a more inclusive approach and broader discussions on governance.

Ademola Jegede underscores the need to address the implications of GE for a wide range of human rights in developing countries in his chapter "Who will bell the cat? African Union, Geoengineering and the potential role of human rights". The chapter departs from the African Charter on Human and Peoples' Rights, supported by other policy and legal instruments that reflect the commitment of states to addressing climate change. While there is little research on GE at the African Union level, Jegede emphasizes its importance in assessing the adequacy of regional legal and policy frameworks. The chapter argues that there are opportunities and challenges in an Africa-wide implementation of afforestation, BECCS, and ocean fertilization that may both enhance and threaten key human rights —in particular, the right to life, the right to have access to sufficient food and water, the right to have access to health care services, the right to property, the right to adequate housing, and the right to environment. These rights are guaranteed under the African Charter as complemented by potentials in other African Union legal and policy documents relevant to GE.

The special contribution "Reflections on GE and Biodiversity", written by *Perses Bilimoria*, delves into the intricate relationship among biodiversity, GE, and human activities in the face of escalating global warming. It highlights the pivotal role each species plays in maintaining the delicate balance of ecosystems. The chapter discusses GE as a method to restore Earth's natural ecosystems, which are under unprecedented strain due to human-induced warming. The chapter underscores the urgency of the situation, with global warming levels consistently exceeding 1.5 °C. Bilimoria argues for the application of innovative yet largely unvalidated technologies, such as CDR and SRM, to guide us toward a sustainable future. He concludes by emphasizing the complex interplay between biodiversity, GE, and human actions in shaping our future.

In the chapter "Geoengineering and International Law", *Alla Pozdnakova* emphasizes the importance of international law in governing GE activities. Despite existing sectoral regulations and general principles, ambiguities persist: these uncertainties encompass the legality of GE activities, the application of environmental law principles, and the due diligence standards for states. The chapter suggests that these ambiguities may result in conflicts, particularly concerning the transboundary impacts of GE activities. Consequently, Pozdnakova advocates for more comprehensive and specific international regulations and standards tailored to GE activities.

In her chapter "Geoengineering and Global Commons-What Role for Ecocide Law?", *Anna Maddrick* examines the concepts of solar GE and ecocide law. She argues that despite the theoretical nature of both wide-scale solar GE deployment and the implementation of a crime of ecocide, there are many parallels between the two phenomena. The chapter begins by briefly reviewing the definitional and historical parallels between the concepts, including their focus on scale and intent. Maddrick suggests that ecocide law could naturally build upon existing international law and serve as a useful tool for analyzing the legality of solar GE deployment. Furthermore, she posits that ecocide law, in conjunction with polycentric governance, could provide more robust environmental protection for the global commons in the face of solar GE. This approach could also offer a legitimate and robust legal framework for discussing such developments, reinforced by internationally enforceable rights on Earth and in outer space.

In their chapter "Geoengineering and Outer Space", *Anthi Koskina* and *Manolis Plionis* explore space-based GE, which involves the benevolent manipulation of the planetary environment from space. The chapter distinguishes between GE technologies that directly alter the environment's composition and those that utilize novel, untested technologies, such as space-based solar shields, for the same purpose. Koskina and Plionis emphasize that if such a path is pursued, these technologies must be carefully regulated in legal terms to ensure compliance with the Outer Space Treaty (1967), particularly Article IX, which aims to protect the natural environment. However, they also acknowledge that the unfamiliarity with this new type of activity may necessitate a review of existing legislation and even a consideration of new approaches rather than strictly adhering to the existing rules of international space law.

Rick Wylie explores in his chapter "Geoengineering and Public Value" the concept of public value in the context of GE. The chapter unpacks the potential contributions of public value to the GE policy debate, addressing key issues associated with the implementation of GE, specifically focusing on SRM. It acknowledges the nascent stage of GE implementation and the uncertainties that surround its deployment and impact. The chapter underscores the need for an anticipatory, agile governance model that can evolve as GE progresses politically, scientifically, and publicly. It also highlights the current absence of a dedicated international framework for SRM governance. Wylie advocates for a public value approach to inform the development of a governance framework for GE, bridging the gap between the micro-level experiences of individual citizens and the macro-level societal and public sphere.

The chapter "Public Engagement with Geoengineering" by *António Carvalho* delves into the current efforts to involve the public in GE discussions. It scrutinizes how the experimental and contentious nature of these technologies influences participatory methodologies. A comprehensive review of research literature from the past 15 years is conducted, including focus groups, surveys, and innovative citizen engagement practices. The chapter probes into various participatory strategies, moving beyond the mere identification of pre-established ethical concerns. Empirical data from unique focus groups on SRM and CDR in Coimbra, a first in Portugal, is analyzed. These groups, designed with an emphasis on performance, experimentation, and speculation, aim to reveal how participants' ideologies, politics, and cultures shape Geoethics in the context of the ongoing climate crisis. Carvalho highlights the experimental potential of GE, arguing that these technologies provide a platform for novel approaches to public engagement with science and technology. The chapter underscores the crucial role of social sciences, environmental humanities, and participatory methods in shaping collective visions of a geoengineered future.

In his chapter "GE Extended: Terraforming on Celestial Bodies", *Chris Pak* employs science fiction (sf) as a tool for analyzing and critiquing terraforming. He contextualizes theoretical and speculative ideas within broader social, historical, and conceptual frameworks, and narrates the outcomes of specific decisions or frameworks related to the adaptation and colonization of celestial bodies. This genre of sf serves as an experimental platform for formulating and scrutinizing strategies for projects of planetary scale. Terraforming sf engages in debates about deliberation and the potential for developing international decision-making and coordination frameworks under crisis conditions. The chapter explores a variety of sf works, including Andy Weir's "Project Hail Mary", Mary Robinette Kowal's "Lady Astronaut" trilogy, and Kim Stanley Robinson's "Mars" trilogy, among others. These works are used to contemplate the impact of catastrophes on decision-making, the emergence of new institutions enabling terraforming, the corporate exploitation of space, and the significance of forming new identities for the survival of interplanetary colonies.

8 Facing the "risk vs. risk" when Turning Earth's Thermostat down

Addressing the complex relationship between climate action and GE involves a nuanced consideration of risks. In grappling with the intricate challenge of GE, specifically the endeavor of reducing Earth's temperature, we find ourselves in a profound "risk vs. risk" predicament. The complex interplay of environmental, political, economic, and ethical factors presents a myriad of risks that require meticulous consideration. When balancing the potential hazards of climate change against the uncertainties tied to the implementation of GE technologies, a detailed comprehension of the interconnected systems involved is crucial. This includes understanding the concept of "risk tipping points" and the potentially catastrophic impacts when these points are crossed. The pressing need to address climate change, underscored by the escalating rate of global warming, compels us to explore innovative solutions. Yet, the act of interfering with Earth's natural systems introduces a range of risks that extend across ecological, geopolitical, and societal realms. One such risk is the potential disruption of monsoon patterns that could affect agricultural systems across the world; an example of how geoengineering interventions could push us toward multiple risk tipping points.

This “risk vs. risk” situation highlights the necessity for a balanced and interdisciplinary approach that acknowledges the fundamental uncertainties involved. Scientific research, guided by ethical guidelines and robust governance structures, forms the foundation of responsible GE investigation. International cooperation, learning from governance challenges in other emerging technologies, can aid in developing consensus and regulatory mechanisms. However, we must also confront the possibility that traditional approaches to managing technological risks may be inadequate for planetary-scale interventions that could fundamentally alter Earth’s systems. As we contemplate the implementation of GE technologies, it is essential to learn from history, taking into account the unintended repercussions of previous technological interventions. The societal and public aspects, as highlighted by public opinions and attitudes, stress the importance of transparent communication and involvement.

Philosophical viewpoints, resonating with the principles of existentialism, utilitarianism, and deontology, remind us of the ethical and moral obligations intertwined with modifying Earth’s systems (Hamilton, 2013; Preston, 2013). As we navigate the intricate “risk vs. risk” landscape, the insights of Slavoj Žižek, as presented in his May 2017 lecture and subsequent book ‘The Courage of Hopelessness’ (2017), prompt us to confront the inherent ideological hurdles and biases shaping our risk perceptions and the possible consequences of GE measures. Žižek encourages us to face the “courage of hopelessness”, an insight which is especially pertinent for our historical moment when even the most pessimist diagnostics as a rule finishes with an uplifting hint at some version of the proverbial light at the end of the tunnel. He argues that true courage is not to imagine an alternative, but to accept the consequences of the fact that there is no discernible alternative. This involves questioning our assumptions, broadening our perspectives, and striving for a more nuanced understanding of the complex interplay of the risks involved in GE.

The current state of public discourse around geoengineering appears to maintain a generally cautious orientation, though this balance may shift as deployment discussions intensify and various stakeholders seek to influence public opinion through information campaigns.

9 A Framework for Utilizing GE for Practitioners

Drawing on the insights from our contributors and the risk vs risk approach discussed earlier, we set out the following ten steps to represent a streamlined framework for the implementation of geoengineering by practitioners. In addition, we made the test with ChatGPT asking for “Ten Steps for Responsible Geoengineering”. The result came close to our deliberations showing that the web already contains a good amount of information for a substantive assessment of this kind. It showed even further that—contrary to other areas where there obviously is a bias for an industrial countries angle like in the field of the international information order—the pendulum of the information basis for GE is (still) swinging in the direction of a cautionary approach. But this might change, the closer we get to implementation, when information on this will flood the networks. Now, in the following, our settled version for the "10 Steps for Responsible Geoengineering":

1. Research and Assessment: Begin by conducting thorough research on various geoengineering techniques, their potential benefits, risks, and ethical implications.
2. Identify Goals: Define specific climate-related goals and objectives that geoengineering interventions aim to address, such as reducing global temperature or mitigating the impacts of extreme weather events.
3. Select Suitable Techniques: Choose appropriate geoengineering methods based on their feasibility, effectiveness, and potential side effects. Examples include solar radiation modification or carbon dioxide removal techniques.
4. Modeling and Simulation: Utilize climate models and simulations to predict the outcomes of implementing different geoengineering strategies under various scenarios.
5. Risk Assessment: Assess the potential environmental, social, and geopolitical risks associated with each geoengineering approach, considering factors like unintended consequences, regional disparities, and ethical concerns.
6. International Cooperation: Engage in international collaboration and dialogue to ensure transparency, cooperation, and shared responsibility in geoengineering research and implementation.
7. Regulatory Framework: Establish regulatory frameworks and governance mechanisms at local, national, and international levels to oversee geoengineering activities, address concerns, and minimize risks.
8. Public Engagement: Conduct extensive public outreach and education to increase awareness, gather feedback, and address public concerns and perceptions about geoengineering.
9. Pilot Projects: Initiate small-scale pilot projects or field experiments to test the feasibility, efficacy, and safety of selected geoengineering techniques in real-world conditions while closely monitoring their impacts.
10. Monitoring and Adaptation: Continuously monitor and evaluate the effectiveness, side effects, and unintended consequences of implemented geoengineering interventions. Adjust strategies as needed based on scientific findings and stakeholder feedback.

It is important to note that implementing geoengineering is a highly complex and controversial endeavor, and careful consideration must be given to ethical, social, and environmental implications, as well as the potential for unintended consequences. It is crucial to recognize that geoengineering complements but does not replace mitigation and adaptation efforts aimed at tackling the underlying causes of climate change.

10 What Impact Do we Expect from our Book

This book stands apart from the existing scientific literature on GE and the reports by academic institutions and think tanks. The contributors aim to provide an informed debate on the politics of GE, covering all aspects—environmental, political, and economic, legal, and regulatory, societal, and public. The goal is to offer a comprehensive outlook on how a research programme for the politics of GE should be structured. In the course of the preparation of this book, the work and reflection benefitted from the opportunity to organize a virtual session with some of our contributors at the Science Summit at the 78th United Nations General Assembly on 27 September 2023, as well as other occasions (Fig. 1). For instance, intermediary results were presented and discussed as at the George C. Marshall European Center for Security Studies on 28 July 2023 (Bastian et al., 2023). These inputs considerably helped to shape this publication given it becoming a useful tool also for practical solutions and management approaches.

This book signifies a collaborative global initiative committed to comprehensively addressing the intricate facets of GE. Contributors, including academics and practitioners from diverse institutions and countries, provide unique disciplinary perspectives in their chapters. The editorial process involved a continuous dialogue, weaving

Fig. 1 Contributors to the session "Turning the thermostat down—but how? Towards a Research Agenda for the Politics of Geoengineering" at the Science Summit during the 78th United Nations General Assem bly (27 September 2023). The panel brought together Kai-Uwe Schrogl (European Space Agency), Ntorina Antoni (Eindhoven University of Technology), Ademola Oluborode Jegede (University of Venda), Rick Wylie (University of Central Lancashire), Anna Maddrick (Stop Ecocide International), Joshua Horton (Harvard University), Alla Pozdnakova (University of Oslo) and Perses Bilimoria (Climate Change Due Diligence Group) to discuss research priorities, governance, and societal perspectives on geoengineering

theory contributions with practical implications during extensive review cycles. Authors, serving as mutual peer-reviewers, enhanced the discourse. This collaborative effort facilitated cross-referencing and validation across disciplines examining different components of GE policy. The three co-editors extend their sincere appreciation to all contributors for their commitment to this volume, reflecting a blend of unique insights. Thanks also to Susanna Gigante and Sjarien Smit in contributing to editing this manuscript. We very much appreciate the readiness of Annett Buettner and Johannes Glaser from Springer and to the series editor Professor John Dodson to include this publication in the Springer Climate series. This endeavor aims to be a sophisticated, up-to-date resource, assisting members of the global community, academic audiences, and other stakeholders in navigating the current and future trajectories of GE. The authors acknowledge the use of large language models for literature review assistance and citation formatting. All content analysis, arguments, and conclusions represent the authors' independent scholarly work.

11 GE Timeline

Alternative Text for Accessibility: A chronological table spanning 1923–2024 showing key developments in geoengineering history, from early conceptual work through recent policy developments including international governance efforts and the UK's major funding commitment.

Year	Event
1923	Oberth, H. *Die Rakete zu den Planetenräumen*–introduced the concept of space mirrors.
1972	Convention on the prevention of marine pollution by dumping of wastes and other matter.
1977	Environmental modification (ENMOD) convention–prohibits military or hostile uses of environmental modification techniques.
1980s–1990s	Crutzen, Paul J.–work on the ozone layer and the Anthropocene concept.
1985	Vienna convention for the protection of the ozone layer–framework for international efforts to protect the ozone layer.
1987	Montreal protocol on substances that deplete the ozone layer–agreement to phase out ozone-depleting substances.
1992	United Nations framework convention on climate change (UNFCCC)–international treaty to combat climate change.
1996	London protocol to the convention on the prevention of marine pollution by dumping of wastes and other matter.
1997	Kyoto protocol–commits state parties to reduce greenhouse gas emissions.
1998	Energy charter treaty–establishes a multilateral framework for cross-border cooperation in the energy industry.

(continued)

(continued)

Year	Event
2005	International Committee of the red Cross—Environmental modification convention.
2006	Crutzen publishes "albedo enhancement by stratospheric Sulfur injections"–pivotal paper legitimizing geoengineering research.
2009	Royal Society–*geoengineering the climate: Science, governance, and uncertainty.*
2009	Amendment to the London protocol–facilitates CO2 export for sub-seabed storage.
2010	Oxford Martin School–*Oxford principles for geoengineering governance.*
2010	The World Academy of Sciences, Royal Society, environmental Defense fund–geoengineering methods.
2010	Aichi biodiversity targets under the Convention on Biological Diversity—includes a moratorium on geoengineering.
2011	Stanford University–*assessing the feasibility of global warming mitigation measures.*
2011	Singer–ethical implications of geoengineering.
2013	ETC group–governance frameworks for geoengineering.
2013	Preston–solar geoengineering controversy.
2013	Rules for ocean fertilization under the 1996 London protocol—Prohibits commercial ocean fertilization activities.
2014	IPCC–Annex III: Glossary in *Climate Change 2014: Synthesis Report. Definition of geoengineering.*
2014	Scheer & Renn–Public engagement in geoengineering.
2015	Armeni & Redgwell–Legal framework for geoengineering.
2015	Kahan et al. –public perception of geoengineering.
2015	National Academy of Sciences–*climate intervention: Reflecting sunlight to cool earth.*
2015	Paris agreement–legally binding international treaty aiming to limit global warming to well below 2 °C.
2015	China begins building world's largest federally funded geoengineering research program.
2016	Schrogl, K.-U. and Summerer, L.–*Climate engineering and space, Acta Astronautica (129), pp. 121–129.*
2018	Rahman, Artaxo, Asrat, & Parker—Regional workshops on geoengineering.
2019	National Academy of Sciences—*Negative emissions technologies and reliable sequestration.*
2019	China completes major federally funded geoengineering research phase.
2020	International risk governance Center (IRGC) publishes policy recommendations for international GE regulatory framework.
2020	NOAA begins receiving congressional funding for geoengineering research.

(continued)

(continued)

Year	Event
2022	Over 500 scientists from 60+ countries endorse solar geoengineering non-use agreement.
2022	Kunming-Montreal global biodiversity framework aims to halt biodiversity loss by 2030.
2023	COP28 Dubai achieves historic agreement to "transition away from fossil fuels".
2023	Center for International Environmental law–Statement on geoengineering UN general assembly requests ICJ advisory opinion on climate obligations (March 29).
2023	Mexico becomes first country to ban outdoor solar geoengineering experiments (January).
2023	Alliance for just deliberation on solar geoengineering established (April).
2023	White house publishes first comprehensive report on solar radiation modification (June).
2023	EU Commission characterizes SRM as representing "unacceptable level of risk" (June).
2023	Climate overshoot commission recommends SRM research with deployment moratorium (September).
2023	Germany adopts climate foreign policy strategy opposing SRM development (December).
2023	Switzerland ratifies 2009 amendment to London protocol for CO2 export (November).
2024	Switzerland's UN resolution on SRM governance fails at UNEA-6 (February).
2024	Harvard permanently halts SCoPEx stratospheric experiment (March 18).
2024	David Keith relocates to University of Chicago to establish climate systems engineering initiative.
2024	UK ARIA announces £56.8 million funding for 21 climate cooling projects–largest single source of solar geoengineering research funding globally.
2024	Major philanthropic commitments exceed $100 million (quadrature climate foundation $40 M, Simons Foundation $50 M).
2024	EU AI act enters into force providing potential regulatory model for emerging technologies (August).
2024	EU Group of chief scientific advisors releases SRM recommendations advocating deployment moratorium (December).
2024	ICJ hearings on state climate obligations conclude with unprecedented 96 states participating (December 2–13).
Target by 2025	China plans weather modification system covering 5.5 million square kilometers.
Target by 2025	ICJ advisory opinion on state climate obligations.

References

Academic Sources

Peer-Reviewed Journal Articles

Babones, S. (2020). From Tianxia to Tianxia: The return of history in Chinese international thought. *Chinese Political Science Review, 5*(2), 131–140.

Bastian, K., Cantafio Apitz, C., Hagemann, F., & Tettweiler, F. (2023). Translating the climate-security nexus. *George C. Marshall Center Policy Brief.* https://www.marshallcenter.org/en/publications/security-insights/policy-brief-translating-climate-security-nexus-0

Biermann, F., Oomen, J., Gupta, A., Ali, S. H., Conca, K., Hajer, M. A.,... & VanDeveer, S. D. (2022). Solar geoengineering: The case for an international non-use agreement. *Wiley Interdisciplinary Reviews: Climate Change, 13*(3), e754.

Boettcher, M., & Schäfer, S. (2017). Reflecting upon 10 years of geoengineering research: Introduction to the Crutzen+ 10 special issue. *Earth's Future, 5*(3), 266–277.

Buck, H. J., Gammon, A. R., & Preston, C. J. (2014). Gender and geoengineering. *Hypatia, 29*(3), 651–669.

Carton, W. (2019). "Fixing" climate change by mortgaging the future: Negative emissions, spatiotemporal fixes, and the political economy of delay. *Antipode, 51*(3), 750–769.

Crutzen, P. J. (2006). Albedo enhancement by stratospheric sulfur injections: A contribution to resolve a policy dilemma? *Climatic Change, 77*(3), 211–220.

Fagerberg, J., Mowery, D. C., & Nelson, R. R. (Eds.). (2005). *The Oxford handbook of innovation.* Oxford University Press.

Floridi, L., Cowls, J., Beltrametti, M., Chatila, R., Chazerand, P., Dignum, V., & Vayena, E. (2018). AI4People—An ethical framework for a good AI society: Opportunities, risks, principles, and recommendations. *Minds and Machines, 28*, 689–707.

Kahan, D. M., Jenkins-Smith, H., Tarantola, T., Silva, C. L., & Braman, D. (2015). Geoengineering and climate change polarization: Testing a two-channel model of science communication. *The Annals of the American Academy of Political and Social Science, 658*(1), 192–222.

Keohane, R. O., & Nye, J. S., Jr. (1973). Power and interdependence. *Survival, 15*(4), 158–165.

Markusson, N., McLaren, D., & Tyfield, D. (2018). Towards a cultural political economy of mitigation deterrence by negative emissions technologies (NETs). *Global Sustainability, 1*, e10.

McEvoy, D. M., McGinty, M., Cherry, T. L., & Kroll, S. (2024). International climate agreements under the threat of solar geoengineering. *Journal of the Association of Environmental and Resource Economists, 11*(4), 853–886.

Mclaren, D., & Corry, O. (2025). Solar geoengineering research faces geopolitical deadlock. *Science. 387(6729), 28-30.*

Möller, I. (2020). Political perspectives on geoengineering: Navigating problem definition and institutional fit. *Global Environmental Politics, 20*(2), 57–82.

Moreno-Cruz, J. B., & Smulders, S. (2017). Revisiting the economics of climate change: The role of geoengineering. *Research in Economics, 71*(2), 212–224.

Preston, C. J. (2013). Ethics and geoengineering: Reviewing the moral issues raised by solar radiation management and carbon dioxide removal. *Wiley Interdisciplinary Reviews: Climate Change, 4*(1), 23–37.

Rahman, A. A., Artaxo, P., Asrat, A., & Parker, A. (2018). Developing countries must lead on solar geoengineering research. *Nature, 556*(7699), 22–24.

Ramos, Y., & Santos, F. D. (2025). Are climate geoengineering technologies being patented? An overview. *Climate, 13*(4), 77.

Rayner, S., Heyward, C., Kruger, T., Pidgeon, N., Redgwell, C., & Savulescu, J. (2013). The oxford principles. *Climatic Change, 121*(3), 499–512.

Sartre, J. P. (2018). Being and nothingness: An essay in phenomenological ontology (S. Richmond, Trans.). Routledge. (Original work published 1943)
Scheer, D., & Renn, O. (2014). Public perception of geoengineering and its consequences for public debate. *Climatic Change, 125*, 305–318.
Schrogl, K. U. (2022). Mit Geoengineering gegen den Klimawandel. *Internationale Politik, 77*(6), 92–97.
Schrogl, K. U., & Summerer, L. (2016). Climate engineering and space. *Acta Astronautica, 129*, 121–129.
Stephens, J. C., & Surprise, K. (2020). The hidden injustices of advancing solar geoengineering research. *Global Sustainability, 3*, e2.
Talberg, A., Christoff, P., Thomas, S., & Karoly, D. (2018). Geoengineering governance-by-default: An earth system governance perspective. *International Environmental Agreements: Politics, Law and Economics, 18*, 229–253.
Temple, J. (2024b, June 14). This London nonprofit is now one of the biggest backers of geoengineering research. *MIT Technology Review*. https://www.technologyreview.com/2024/06/14/1093778/foundations-are-lining-up-to-fund-geoengineering-research/
The State Council of PRC. (2020). China to forge ahead with weather modification service. *The State Council of PRC*. https://english.www.gov.cn/policies/latestreleases/202012/02/content_WS5fc76218c6d0f7257694125e.html
Whyte, K. (2018). Indigenous science (fiction) for the Anthropocene: Ancestral dystopias and fantasies of climate change crises. *Environment and Planning E: Nature and Space, 1*(1–2), 224–242.

Academic Books and Book Chapters

Björkdahl, A., & Selimovic, J. M. (2019). WPS and civil society. In *The Oxford Handbook of Women, Peace, and Security* (pp. 428–438). Oxford University Press.
Callies, D. E. (2023). The ethics of geoengineering. In *Handbook of the philosophy of climate change* (pp. 919–937). Springer International Publishing.
Chesbrough, H. W. (2003). *Open innovation: The new imperative for creating and profiting from technology*. Harvard Business School.
Christensen, C. M. (1997). *The innovator's dilemma: When new technologies cause great firms to fail*. Harvard Business School Press.
Hamilton, C. (2013). *Earthmasters: The dawn of the age of climate engineering*. Yale University Press.
Kahneman, D., & Tversky, A. (2013). Prospect theory: An analysis of decision under risk. In L. C. MacLean & W. T. Ziemba (Eds.), *Handbook of the fundamentals of financial decision making: Part I* (pp. 99–127). World Scientific Publishing.
Kant, I. (1785). *Groundwork for the Metaphysics of Morals*. (Original work published 1785).
Krugman, P. (2009). *The return of depression economics and the crisis of 2008*. Norton & Company.
Mazzucato, M. (2021). *Mission economy: A moonshot guide to changing capitalism*. Penguin UK.
Mbembe, A. (2019). *Necropolitics* (S. Corcoran, Trans.). Duke University Press. (Original work published 2011).
Mearsheimer, J. J. (2003). *The tragedy of great power politics* (Updated edition). Norton & Company.
Mill, J. S. (1863). *Utilitarianism*. Parker, Son & Bourn.
Oberth, H. (1923). *Die Rakete zu den Planetenräumen*. R. Oldenbourg.
Rogers, E. M., Singhal, A., & Quinlan, M. M. (2014). Diffusion of innovations. In *An integrated approach to communication theory and research* (pp. 432–448). Routledge.

Scherrer, C., Garcia, A., & Wullweber, J. (2023). Introduction to the handbook on critical political economy and public policy. In *Handbook on critical political economy and public policy* (pp. 1–18). Edward Elgar Publishing.
Singer, P. (2011). *The expanding circle: Ethics, evolution, and moral progress.* Princeton University Press.
Wendt, A. (1999). *Social theory of international politics* (Vol. 67). Cambridge University Press.
Zhao, T. (2021). *All under heaven: The Tianxia system for a possible world order* (J. Cayley & J. Guo, Trans). University of California Press. (Original work published 2005).
Žižek, S. (2017). *The courage of hopelessness: Chronicles of a year of acting dangerously.* Penguin UK.

Reports and Policy Documents

AI Now Institute. (2023). Algorithmic Accountability: Moving Beyond Audits. 11 April, 2023. https://ainowinstitute.org/publication/algorithmic-accountability
Armeni, C., & Redgwell, C. (2015). *International legal and regulatory issues of climate geoengineering governance: rethinking the approach.* Climate Geoengineering Governance Working Paper Series, 21, 6–8.
Carnegie Climate Governance Initiative. (2023, September 14). *Global status of activities relating to Solar Radiation Modification and its governance* (9th Ed.). https://www.c2g2.net/wp-content/uploads/20230914-C2GSRMGlobalStatusUpdate.pdf
Center for International Environmental Law (CIEL). (2023). Statement on Geoengineering Moratorium. https://www.ciel.org/news/cbd-cop16-reaffirms-geoengineering-moratorium-fails-fossil-fuel-phaseout/
Climate Overshoot Commission. (2023). Report on Solar Radiation Modification. https://www.overshootcommission.org/report
ETC Group. (2013). *The Artificial Intelligence of Geoengineering.* Communiqué. https://www.etcgroup.org/content/artificial-intelligence-geoengineering
Frontier Climate. (2023, September 7). *Fall 2023 purchases: Facilitating third round of carbon removal purchases.* https://frontierclimate.com/writing/fall-2023-purchases
Geoengineering Monitor. (2021). *Ocean fertilization. Geoengineering Technology Briefing.* https://www.geoengineeringmonitor.org/wp-content/uploads/2021/04/ocean-fertilization.pdf
Geoengineering Monitor. (2024). *Hands Off Mother Earth Alliance and global resistance networks.* https://www.geoengineeringmonitor.org/2024/04/hands-off-mother-earth-alliance/
GOV.UK. (2024). Drax bioenergy with carbon capture and storage project development consent decision announced. *GOV.UK*, https://www.gov.uk/government/news/drax-bioenergy-with-carbon-capture-and-storage-project-development-consent-decision-announced
Hands Off Mother Earth Campaign. (2022). *Hands Off Mother Earth manifesto against geoengineering.* https://www.geoengineeringmonitor.org/hands-off-mother-earth-manifesto-against-geoengineering
Heinrich Böll Stiftung. (2019). *High-Emitting, Oil-Producing Countries Block Progress on Geoengineering Governance at United Nations.* https://www.boell.de/en/2019/03/14/civil-society-reiterates-call-ban-geoengineering
IEA. (2022). *Direct air capture: A key technology for net zero.* International Energy Agency.
Kim, J., & Boulanin, V. (2023). *AI for climate security: An assessment of machine learning's role in climate and security challenges.* SIPRI. https://www.sipri.org/sites/default/files/2023-12/2312_sipri_policy_report_ai_for_climate_security_1.pdf
Moore, M., & Alexander, L. (2007). Deontological ethics. Edward N. Zalta. Stanford Encyclopedia of Philosophy (en inglés) (Fall 2008 Edition).
National Academies of Sciences, Engineering, and Medicine (NASEM). (2019). Negative emissions technologies and reliable sequestration: A research agenda. *The National Academies*

Press. https://nap.nationalacademies.org/catalog/25259/negative-emissions-technologies-and-reliable-sequestration-a-research-agenda

National Academies of Sciences, Engineering, and Medicine. (2021). *Reflecting sunlight: Recommendations for solar geoengineering research and research governance.* The National Academies Press. https://doi.org/10.17226/25762

National Research Council. (2015). *Climate intervention: Reflecting sunlight to cool earth.* https://doi.org/10.17226/18988

OECD. (2024). *Framework for anticipatory governance of emerging technologies.* https://www.oecd.org/en/publications/2024/04/framework-for-anticipatory-governance-of-emerging-technologies_14bf0402.html

Royal Society. (2009). *Geoengineering the climate: Science, governance and uncertainty (RS Policy Document 10/09).The Royal Society.* https://royalsociety.org/topics-policy/projects/geoengineering-climate/report/

UN Climate Change. (2013). *COP28 agreement signals "beginning of the end" of the fossil fuel era.* UNClimateChange_COP28_FossilFuelEra_2023

United Nations Environment Programme (UNEP). (2023). *One Atmosphere: An Independent Expert Review on Solar Radiation Modification Research and Deployment.* https://www.unep.org/resources/report/Solar-Radiation-Modification-research-deployment

World Economic Forum (WEF). (2024). *Global Risks Report 2024.* https://www.weforum.org/publications/global-risks-report-2024/

Zero Emissions Platform. (2024). *The industrial carbon management strategy: A landmark recognition for CCS and CCU.* https://zeroemissionsplatform.eu/publication/the-industrial-carbon-management-strategy-a-landmark-recognition-for-ccs-and-ccu/

Primary Legal and Policy Sources

International Treaties and Conventions

Treaty on Principles Governing the Activities of States in the Exploration and Use of Outer Space, Including the Moon and Other Celestial Bodies (Outer Space Treaty). (1967, January 27). Opened for signature at London, Moscow, and Washington. Entered into force October 10, 1967. 610 U.N.T.S. 205; 18 U.S.T. 2410.

Government Policy Documents

Ministry of Climate, Energy and Utilities. (2023, November 27). Aalborg Declaration on enabling cross-border carbon capture, utilisation and storage (CCUS) in Europe. *Danish Ministry of Climate, Energy and Utilities.* https://www.kefm.dk/Media/638641513877792088/Aalborg%20declaration.pdf

Auswärtiges Amt. (2023). Climate Foreign Policy Strategy. https://www.auswaertiges-amt.de/resource/blob/2633116/8ea71a517304862954bff70434c3994e/kap-strategie-en-data.pdf

European Commission. (2023). *Joint communication to the European Parliament and the Council: A new outlook on the climate and security nexus*: Addressing the impact of climate change and environmental degradation on peace, security and defence (JOIN/2023/19 final). https://eur-lex.europa.eu/legal-content/EN/TXT/?uri=CELEX:52023JC0019

European Commission. (2024). *Communication from the Commission to the European Parliament, the Council, the European Economic and Social Committee and the Committee of the Regions*: Towards an ambitious Industrial Carbon Management for the EU (COM/2024/62 final). https://eur-lex.europa.eu/legal-content/EN/TXT/?uri=CELEX:52024DC0062

European Commission, Group of Chief Scientific Advisors. (2023). *Scoping paper: Solar radiation modification*. https://research-and-innovation.ec.europa.eu/system/files/2023-08/Scoping_paper_SRM.pdf

European Commission, Group of Chief Scientific Advisors. (2024). Solar Radiation Modification. Scientific Opinion No. 17. https://scientificadvice.eu/advice/solar-radiation-modification/

European Parliament and Council of the European Union. (2024). *Regulation (EU) 2024/1689 of the European Parliament and of the Council of 13 June 2024 laying down harmonised rules on artificial intelligence (AI Act)*. https://eur-lex.europa.eu/eli/reg/2024/1689/oj/eng

European Parliament. (2023, December 8). *EU withdrawal from the Energy Charter Treaty (EPRS Briefing No. 754.632)*. European Parliament. https://www.europarl.europa.eu/RegData/etudes/BRIE/2023/754632/EPRS_BRI(2023)754632_EN.pdf

German Environment Agency. (2023). Report on Carbon Capture and Storage. https://www.umweltbundesamt.de/sites/default/files/medien/11850/publikationen/ccs_in_the_global_climate_debate_policy_brief.pdf

Science Summit at UNGA78. (2023). [VIRTUAL] Turning the thermostat down – but how? Towards a research agenda for the politics of geoengineering (Session 270906).,*Science Summit at UNGA78. (2023, September 27)*. https://sciencesummitunga78.sched.com/event/1O4if/virtual-turning-the-thermostat-down-but-how-towards-a-research-agenda-for-the-politics-of-geoengineering-270906

Swiss Federal Council. (2023). Federal Council lays groundwork for export of CO2 for sub-seabed storage. https://www.admin.ch/gov/en/start/documentation/media-releases.msg-id-98800.html

The White House. (2023). *Congressionally-mandated report on solar radiation modification*. Office of Science and Technology Policy. https://www.whitehouse.gov/ostp/news-updates/2023/06/30/congressionally-mandated-report-on-solar-radiation-modification/

IPCC Documents

Intergovernmental Panel on Climate Change (IPCC). (2014). *Climate Change 2014: Synthesis Report*. Contribution of Working Groups I, II and III to the Fifth Assessment Report of the Intergovernmental Panel on Climate Change [Core Writing Team, R. K. Pachauri & L. A. Meyer (Eds.)]. IPCC, Geneva, Switzerland, 151. https://www.ipcc.ch/report/ar5/syr/

Intergovernmental Panel on Climate Change (IPCC). (2021). *Climate Change 2021: The Physical Science Basis*. Contribution of Working Group I to the Sixth Assessment Report of the Intergovernmental Panel on Climate Change [V. Masson-Delmotte, P. Zhai, A. Pirani, S. L. Connors, C. Péan, S. Berger, N. Caud, Y. Chen, L. Goldfarb, M. I. Gomis, M. Huang, K. Leitzell, E. Lonnoy, J. B. R. Matthews, T. K. Maycock, T. Waterfield, O. Yelekçi, R. Yu, & B. Zhou (Eds.)]. Cambridge University Press. https://www.ipcc.ch/report/ar6/wg1/

Intergovernmental Panel on Climate Change (IPCC). (2023). *Climate change 2023: Synthesis report*. Contribution of Working Groups I, II and III to the Sixth Assessment Report of the Intergovernmental Panel on Climate Change (Core Writing Team, H. Lee & J. Romero, Eds.). https://www.ipcc.ch/report/ar6/syr/downloads/report/IPCC_AR6_SYR_FullVolume.pdf

International Communications and Declarations

International Committee of the Red Cross. (2005). China: Convention on the prohibition of military or any hostile use of environmental modification techniques (ENMOD), 1976 declaration. https://ihl-databases.icrc.org/en/ihl-treaties/enmod-1976/state-parties/cn

Open Letter on Solar Geoengineering Non-Use Agreement. (2022). https://www.solargeoeng.org/non-use-agreement/

UNFCCC. (2023). COP28 agreement signals beginning of the end of the fossil fuel era. *UN Climate Change Press Release*. https://unfccc.int/news/cop28-agreement-signals-beginning-of-the-end-of-the-fossil-fuel-era

Secondary Sources

News Articles and Magazine Features

Chalmin, E. (2021). *Norway and the EU: Facing the geopolitics of climate change. Climate diplomacy brief*. adelphi.

Rodriguez, S., & Lo, J. (2023, January 18). Mexico plans to ban solar geoengineering after rogue experiment. *Climate Home News*. https://www.climatechangenews.com/2023/01/18/mexico-plans-to-ban-solar-geoengineering-after-rogue-experiment/

Lo, J. (2024, February 29). Nations fail to agree ban or research on solar geoengineering. *Climate Home News*. https://www.climatechangenews.com/2024/02/29/nations-fail-to-agree-ban-or-research-on-solar-geoengineering-regulations/

De la Garza, A. (2023, February 24). Why billionaires are obsessed with blocking out the sun. *Time*. https://time.com/6258126/solar-geoengineering-billionaries-george-soros/

Drugmand, D. (2023). *Corporate promotion of carbon capture and storage contradicts science, study finds*. DeSmog. https://www.desmog.com/2023/11/30/influencemap-report-carbon-capture-corporations-paris-agreement/

Dunne, D. (2018). Geoengineering carries large risks for the natural world, studies show. *Carbon Brief*. https://www.carbonbrief.org/geoengineering-carries-large-risks-for-natural-world-studies-show/

Jayaram, D. (2019, December 5). Chinas geoengineering build-up poses geopolitical and security risks. *Climate-Diplomacy*. https://climate-diplomacy.org/magazine/cooperation/chinas-geoengineering-build-poses-geopolitical-and-security-risks

Kaufman, R. (2019, March 11). The Risks, Rewards and Possible Ramifications of Geoengineering Earth's Climate. *Smithsonian Magazine*. https://www.smithsonianmag.com/science-nature/risks-rewards-possible-ramifications-geoengineering-earths-climate-180971666/

McLaren, D., & Corry, O. (2024, March 7). Countries failed to agree first steps on geoengineering: What went wrong? *Legal Planet*. https://legal-planet.org/2024/03/07/countries-failed-to-agree-first-steps-on-geoengineering-what-went-wrong/

Scobie, M. (2022, October 12). What is solar radiation modification and what questions should SIDS be asking about the governance of its research and deployment? ODI. https://odi.org/en/insights/what-is-solar-radiation-modification-and-what-questions-should-sids-be-asking-about-the-governance-of-its-research-and-deployment/

Schrogl, K.U., & Wang, G. (2024). #SpaceWatchGL Opinion–Space4Earth–Why space applications become a key to geoengineering and its governance. *Spacewatch Global*. https://spacewatch.global/2024/11/spacewatchgl-opinion-space4earth-why-space-applications-become-a-key-to-geoengineering-and-its-governance/

Sharma, K. (2023, January 31). How China uses geoengineering to pursue a hybrid warfare strategy. *The Diplomat*. https://thediplomat.com/2023/01/how-china-uses-geoengineering-to-pursue-a-hybrid-warfare-strategy/

Temple, J. (2017, August 2). China Builds One of the World's Largest Geoengineering Research Programs. *MIT Technology Review*. https://www.technologyreview.com/2017/08/02/4291/china-builds-one-of-the-worlds-largest-geoengineering-research-programs/

Temple, J. (2022). The US government is developing a solar geoengineering research plan. *MIT Technology Review*, Available at https://www.technologyreview.com/2022/07/01/1055324/the-us-government-is-developing-a-solar-geoengineering-research-plan/

Temple, J. (2024a, March 18). Harvard has halted its long-planned atmospheric geoengineering experiment. *MIT Technology Review*. https://www.technologyreview.com/2024/03/18/1089879/harvard-halts-its-long-planned-atmospheric-geoengineering-experiment/

Organizational Websites and Press Releases

Drax Group. (2024, January 16). *UK Government approves planning application for BECCS at Drax Power Station*. https://www.drax.com/press_release/uk-government-approves-planning-application-for-beccs-at-drax-power-station/

Harvard University. (2024). *Geoengineering Monitoring*. https://geoengineering.environment.harvard.edu/scopex-announcement

Institute for Advanced Sustainability Studies (IASS). (2017). *Climate Engineering Conference Series*. https://sdg.iisd.org/events/climate-engineering-conference-2017-cec17/

Kleinman Center for Energy Policy. (2024). *Alliance for Just Deliberation on Solar Geoengineering*. University of Pennsylvania. https://kleinmanenergy.upenn.edu/research/working-papers/alliance-just-deliberation-solar-geoengineering

North Sea Transition Authority (NSTA). (2023). *Licenses for Carbon Dioxide Storage*. https://www.nstauthority.co.uk/regulatory-information/carbon-storage/carbon-storage-licensing/

Planetary Sunshade Foundation. (2024). Research and Advocacy for Space-Based Solar Radiation Management. https://www.planetarysunshade.org/

SRM360. (2024). *UK ARIA funding for planetary sunshade research*. https://srm360.org/2024/03/uk-aria-funding-planetary-sunshade/

SGDeliberation.org. (2023). What is DSG. *The Alliance for Just Deliberation on Solar Geoengineering*. https://sgdeliberation.org/about/

TWAS. (2018, December 18). Announcing SRM geoengineering research groups in the Global South. https://twas.org/article/announcing-srm-geoengineering-research-groups-global-south

TWAS. (n.d.). Solar Radiation Management Governance Initiative. https://twas.org/solar-radiation-management-governance-initiative

University of Chicago News. (2023). *David Keith joins University of Chicago to establish Climate Systems Engineering initiative*. https://news.uchicago.edu/story/david-keith-joins-university-chicago-establish-climate-systems-engineering

SWOT Analysis for Carbon Dioxide Removal (CDR)

Valeria Pinna, Leonora Burnik, and Laure-Marine Vioujard

Abstract This chapter explores Carbon Dioxide Removal (CDR) as a geoengineering endeavor, focusing on extracting CO2 from the atmosphere for storage in geological, terrestrial, or oceanic reservoirs, or its conversion into products. Through real case studies, it conducts a comprehensive SWOT analysis of CDR techniques, delineating their strengths, weaknesses, opportunities, and threats. By providing a comprehensive examination of CDR approaches and their broader implications, this chapter contributes to advancing our understanding of sustainable climate mitigation strategies in the face of global environmental challenges.

Keywords Carbon Dioxide Removal (CDR) · Carbon capture · Geoengineering · Global warming · Climate Change · SWOT analysis

1 What Is CDR, Overall Analysis, and Elements

CDR (Carbon Dioxide Removal) is one of the two Geoengineering (GE) approaches aimed at addressing climate change, one of humanity's most pressing challenges. With global temperatures on the rise, current climate policies recognize the essential role of CDR in limiting temperature rises. For instance, current global climate policy already assumes a great deal of CDR, since without it, "the achievement of the goal of keeping the increase in global average temperature well below 2°C (3.6°F) seems to be impossible" (Gerrard & Hester, 2018). CDR involves extracting CO2 from the atmosphere, utilizing both natural and technological methods. Natural methods include afforestation/reforestation, soil carbon sequestration, biochar, agroforestry, and blue carbon. Technological approaches include Direct Air Carbon Capture and Storage (DACCS), Bioenergy with Carbon Capture and Storage (BECCS), Ocean Alkalinity Enhancement (OAE), and Carbon Mineralization.

V. Pinna (✉) · L. Burnik · L.-M. Vioujard
European Union Agency for the Space Programme, Prague, Czech Republic

K.-U. Schrogl et al. (eds.), *The Politics of Geoengineering*, Springer Climate,
https://doi.org/10.1007/978-3-032-10095-5_2

CDR does not only aim to mitigate climate change by reducing greenhouse gas concentrations but also to address the root cause of global warming. The scientific community has proposed a wide variety of technologies to accomplish this goal:

- Creating machines that directly remove carbon dioxide from the air (Chandler, 2019);
- Fertilizing the oceans with iron filings or other substances that spur algal growth, which absorbs carbon dioxide before sinking to the bottom (Williamson, Wallace, & others, 2012);
- Creating charcoal that holds the carbon (PNP Charcoal, 2024);
- Growing and burning large quantities of biomass for energy and capturing and sequestering the carbon dioxide (Thangaraj et al., n.d.);
- Exposing certain carbon-absorbing rocks to the air in a process called "enhanced weatherization" (MIT Climate Portal, n.d.).

While a considerable number of CDR techniques have been proposed, further research is still required to fully evaluate their scalability, efficacy, risks, and benefits. Current research, including reports by the Intergovernmental Panel on Climate Change (IPCC), suggests that CDR could play a role in mitigating climate change (IPCC, 2022). On the one hand it could achieve deep, rapid, and sustained emission reductions on its own. On the other hand, it can complement other mitigation methods on global or state level by helping lower net CO2 or net GHG emissions in short term, even out "hard-to-abate" residual emissions and help reach net zero CO2 or net zero GHG emissions in the mid-term or in the long term, if deployed at levels above the annual residual emissions, and thus achieving net negative CO2 or GHG emissions. For instance, the EU-funded project GICO developed by the Italian Telematic University Guglielmo Marconi is developing high-temperature inorganic removal sorbents, CO2 capture sorbents, hydrothermal carbonization, and sorption enhanced gasification with the objective of capturing CO2 and converting it into O2 and producing methanol and electricity (European Commission, 2022b).

Therefore, a consensus is growing among scientists that, in addition to emission reductions, CDR might be needed to avoid catastrophic climate changes. CDR could compensate for a portion of the current shortfall in decarbonization efforts and in the longer term, could help to rectify overshoots and stabilize the current climate situation. For instance, CDR technologies can be useful to offset historical emissions, thereby contributing to climate mitigation strategies (Kriegler et al., 2013). IPCC analysis shows that, despite the current lack of appropriate governance over geoengineering technologies, the deployment of methods such as afforestation and reforestation, and soil carbon sequestration have benefits on food security and biodiversity (IPCC, 2019).

1.1 Natural CDR

Natural CDR methods harness the inherent processes of natural ecosystems to remove CO2 from the atmosphere and sequester it within various components of the environment. These methods capitalize on the ability of ecosystems such as forests, soils, wetlands, and oceans to naturally absorb and store carbon (ClimateSeed, 2024). For example, afforestation and reforestation initiatives involve planting trees or restoring forested areas, which act as carbon sinks by absorbing CO2 during photosynthesis and storing it within their biomass and soil. Agroforestry integrates trees into agricultural landscapes, further enhancing carbon storage while providing additional ecosystem benefits (Nayak et al., n.d.). Similarly, soil carbon sequestration techniques focus on enhancing the capacity of soils to retain carbon through practices like reduced tillage, cover cropping, and organic amendments (Rodrigues et al., 2023). Additionally, biochar production involves converting biomass into a stable form of carbon through pyrolysis, which can then be applied to soils to enhance carbon sequestration (Mulabagal et al., 2022). Lastly, blue carbon ecosystems such as mangroves, seagrasses, and salt marshes efficiently capture and store carbon within their biomass and sediment, contributing to coastal carbon sequestration (World Bank, 2023). These natural CDR methods play a crucial role in mitigating climate change by capturing and storing atmospheric CO2, thereby helping to reduce greenhouse gas concentrations and combat global warming.

1.2 Technological CDR–SWOT Analysis

Technological CDR methods employ advanced technologies to capture and remove CO2 from the atmosphere, often with the aim of storing it in geological formations or converting it into useful products. These approaches complement natural CDR methods by offering scalable solutions to address CO2 emissions from various sources. As shown in Fig. 1, these methods include DACCS, BECCS, OAE and Carbon Mineralization.

2 Direct Air Capture with Carbon Storage (DACCS)

(i) What is it?

DACCS involves extracting CO2 directly from the ambient air using chemical sorbents or filters. DAC technology employs chemical reactions to remove carbon dioxide from the air (World Economic Forum, 2023). The so-captured CO2 can be stored underground, utilized for industrial purposes, or converted into other products through carbon conversion processes. DAC technologies typically rely on chemical reactions or physical processes to selectively capture CO2.

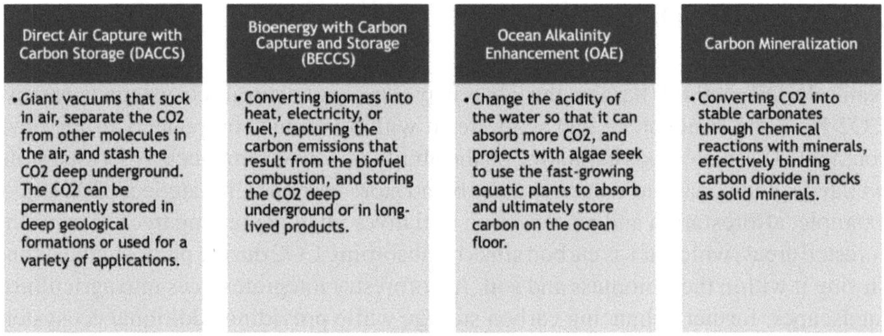

Fig. 1 Overview of technological CDR approaches

(ii) Policy integration and implementation examples.

An exemplary implementation of DAC technology is demonstrated by Orca, a Direct Air Capture and CO2 Storage Plant based in Iceland. The plant was established in 2021 by Climeworks, and it is able to capture 4000 tonnes of CO2 annually (Climeworks, 2021). Climeworks announced its newest and largest DAC and storage plant, Mammoth, on June 28, 2022, aiming for a multi-megaton capacity by 2030 and a gigaton capacity by 2050 (Climeworks, 2022). The technology operates through a three-step process: (1) air intake and filtration, (2) CO2 capture and heating, and (3) CO2 release and storage underground by Carbfix in Iceland. This process ensures permanent carbon removal (Climeworks, 2023).

Moreover, in 2022, the United States announced new funding according to the Inflation Reduction Act (IRA) to drive significant emissions reductions. The IRA aims to create opportunities across the economy, enhancing the establishment of projects, increasing the workforce, and strengthening household energy costs while reducing greenhouse gas emissions (U.S. Department of Energy, 2022). Under the IRA, the US aims at increasing the 45Q tax credit to USD 180/t CO2 captured for storage via DAC, with a capture threshold of as little as 1 kt CO2/year (International Energy Agency, 2022c).

Following the US initiative, the Canadian federal budget proposed an investment tax credit for CCUS projects between 2022 and 2030, valued at around 60% for DAC projects when CO2 is stored at an eligible permanent sequestration site (Canada Revenue Agency, 2024)). According to Canada's Carbon Management Strategy, Canada is well positioned due to its wealth of favorable CO2 storage geology capacity, bringing Canada to a leading position (Natural Resources Canada, 2023). Back in 2010, Alberta developed Canada's first comprehensive carbon sequestration legislation (CanLII, 2010). Reporting some data, "The Alberta government is investing $1.24 billion for up to 15 years in the Quest and Alberta Carbon Trunk Line (ACTL) projects. The Quest and ACTL projects have safely captured and stored a total of more than 11.5 million tonnes of carbon dioxide since 2015" (Canada Energy Regulator, 2022).

In March 2023, the United Kingdom announced the introduction of fundings up to GBP 20 billion for carbon capture, utilization, and storage (CCUS) applications, including for DAC. The idea behind this was to create a competitive carbon capture market by 2023, capable of delivering new job opportunities while supporting the net zero goal (GOV.UK, 2023). The plan, named "the CCUS (Carbon Capture, Usage and Storage) Vision", suggests how the UK government will support the development of early projects to develop CCUS technologies, which aim to store 20–30 million tonnes of CO2 per year by 2030 and support 50,000 jobs by 2030.

Initiatives are also taking place in Europe. On February 6, 2024, the European Commission adopted an Industrial Carbon Management Communication, which underlines and analyzes the contribution provided by carbon capture technologies to the reduction of emissions, estimated to be around 90% by 2040, contributing to reach climate neutrality by 2050. The European Commission aims to store up to 50 MtCO2 a year by 2030, including from DAC (European Commission, 2024).

Finally, starting in June 2023, Japan set a carbon capture roadmap with a target of capturing between 6 and 12 MtCO2 per year by 2030, including from DAC. In particular, the Japan Organization for Metals and Energy Security (JOGMEC) selected seven role model projects with the aim of implementing CCS projects to achieve carbon neutrality by 2050 (Japanese Ministry of Economy, Trade and Industry, 2023).

(iii) Strengths/Opportunities.

DACCS offers numerous benefits, including location independence, as plants can be established wherever CO2 storage facilities exist globally, fostering technological innovation and adaptability (Wesche & Skjølsvold, 2024). Additionally, the precision with which DACCS measures captured CO2 is unparalleled, facilitating meticulous environmental impact assessments and supporting a wide array of applications, from carbon auditing to climate modeling (International Energy Agency Greenhouse Gas R&D Programme, 2021). Furthermore, DAC technology's minimal land footprint offers a stark contrast to more land-intensive methods like afforestation, positioning it as a highly efficient carbon removal strategy (World Resources Institute, 2024). The potential of DACCS is amplified by favorable government policies that incentivize carbon reduction and by robust private investments that propel technological advancements (Inevitable Policy Response, 2023). Among its supporters are the European Commission's Horizon program and visionary private projects such as Microsoft Heirloom, Airbus 1PointFive, and McKinsey Mission Possible Partnership, highlighting the role of public-private partnerships in scaling up climate solutions (Environmental, Social, and Governance Today, 2024).

(iv) Weaknesses/Threats.

Despite its potential, DACCS faces several challenges. According to the International Energy Agency (IEA), direct air capture plants currently operate on a small scale. There are 27 DACCS plants operating worldwide, with the majority capturing CO2 for utilization, such as drinks carbonation. Only two plants store the captured CO2 in geological formations for removal (IEA, 2022b). It is still unclear whether

DACCS plants could be located everywhere in the world. For instance, there are not yet examples of DACCS plants in extremely dry and polluted localities (IEA, 2022b). Despite the presence of private investments, DACCS remains a highly expensive technology. The high cost is primarily due to the significant amount of energy required for the process. DACCS plants need a substantial amount of energy, which ideally should not come from fossil fuel sources to avoid generating additional greenhouse gas emissions (Küng et al., 2023). The energy costs can account for up to 50% of the total operational costs of DACCS plants (IEAGHG, 2021). Other challenges include logistical issues due to the difficulty of finding a perfect location. DACCS plants usually need to be situated in areas with large water storage, given the amount of water needed for the chemical processes involved (Al-Juaied & Whitmore, 2023). Additionally, the deployment of DACCS at a large scale requires strong, long-term policy support and significant financial resources, which are currently lacking (IEA, 2022b)

3 Bioenergy with Carbon Capture and Storage (BECCS)

(i) What is it?

BECCS combines bioenergy production with carbon capture and storage technologies to mitigate carbon dioxide emissions (IEA, 2022a). Biomass, derived from crops, forest residues, or organic waste, is utilized as fuel for energy generation through combustion or conversion processes. The resulting CO2 emissions are then captured and stored underground, preventing their release into the atmosphere. This process effectively removes CO2 from the air while generating renewable energy, offering a dual benefit for climate change mitigation and sustainable energy production (Fajardy & Greenfield, 2024).

ii. Policy integration and implementation examples.

An example of BECCS is offered by the Scottish Carbon Capture & Storage (SCCS) partnership, one of the largest CCS research groups in the UK, consisting of academia, industry, and government. The partnership was established in 2005 between the British Geological Survey, Heriot-Watt University, the University of Aberdeen, the University of Edinburgh, the University of Strathclyde, and other universities across Scotland. The SCCS research is working to accelerate the development of CO2 transportation, capture, and subsurface storage. The group also provides consultancy and impartial advice to industry, the public sector, government agencies, and policymakers (Scottish Carbon Capture & Storage, n.d.). Since its establishment, the SCCS research has held 140 workshops and events and delivered 211 publications and 87 courses. Moreover, the group is also working on establishing partnerships with academia and industry.

Concerning projects, the SCCS is involved in several funded projects on different levels -project management, public engagement, communications and dissemination, and research. Among the financed projects, the CCUS ZEN is exploring the potential for accelerating the deployment of CCUS in two regions in the North Sea region: The Baltic Sea, and the Mediterranean Sea (CCUS ZEN, 2024). The project wants to map emissions sources, storage sites, and utilization options, as well as non-technical mapping connected to stakeholders, regulations, and funding opportunities. The mapping also concerns the identification of relevant users and public or private authorities. Finally, there is the elaboration of detailed plans and business models, as well as exchanging knowledge and know-how. Another project is the HyStorPor, Hydrogen Storage in Porous Media led by University of Edinburgh, which focuses on the large-scale geological storage of energy in the form of hydrogen. The objective of the research is to see whether hydrogen storage in reservoir rocks is feasible. Funded value of the project is £1,117,053 (University of Edinburgh, 2024).

iii. Strengths/Opportunities.

BECCS involves capturing and permanently storing CO2 emitted from biomass processes (Fajardy & Greenfield, 2024). This dual-purpose approach not only reduces atmospheric CO2 levels but also provides a sustainable energy source, addressing both climate change mitigation and energy demands. In addition, its low costs make it a feasible solution as it is reported that per ton of carbon dioxide captured, the cost could be between \$20 and \$100 to \$240 (Vandermel, 2020; Fajardy et al., 2020). BECCS stands out as the only CDR technique capable of simultaneously producing energy. Biogenic sources utilized in BECCS processes can effectively manage emissions from biofuel and biohydrogen production, as well as combustion emissions from various industrial applications, including power plants, waste-to-energy facilities, and biomass co-firing operations in sectors such as cement, pulp, and paper production. Additionally, BECCS can employ biochar as a reducing agent in steel production, further diversifying its applications (Fajardy & Greenfield, 2024). The International Energy Agency recognizes BECCS as a versatile solution with significant potential for carbon removal. Beyond storage, the captured CO2 can be repurposed as a feedstock for manufacturing various products, offering additional value and promoting circular economy principles. While carbon capture and utilization pathways have their advantages, BECCS remains unparalleled in achieving permanent CO2 removal through secure storage and ensuring long-term climate benefits (Fajardy & Greenfield, 2024).

iv. Weaknesses/Threats.

Despite its potential benefits, BECCS faces significant challenges and potential threats that must be carefully addressed for its successful implementation. One major concern is the limited availability of sustainable biomass, which forms the primary feedstock for BECCS. This reliance on biomass may lead to competition with food production, biodiversity conservation efforts, and other essential land uses, raising concerns about environmental and social impacts (Bui et al., 2021; National Academies of Sciences, Engineering, and Medicine, 2018). Moreover, scaling up BECCS operations could exacerbate land-use changes, potentially resulting in deforestation or habitat conversion, thereby endangering biodiversity, soil quality, water resources, and other ecosystem services (American University, 2020; European Commission, 2022a).

Energy efficiency poses another challenge for BECCS, as the process typically incurs energy losses during biomass production, harvesting, transportation, and conversion, leading to reduced overall energy output compared to conventional power plants (Fajardy & MacDowell, 2017). Furthermore, the assumption of carbon neutrality in BECCS operations is critical but uncertain. While BECCS aims to offset carbon emissions from biomass combustion with stored carbon in biomass growth and underground storage, achieving true carbon neutrality depends on various factors, including feedstock sources, land-use changes, and overall carbon balance (Hossain et al., 2023).

Several threats are associated with BECCS, including the risk of underground CO2 storage leakage, which could compromise the long-term effectiveness of carbon removal efforts and pose environmental risks (Fern, 2022). Moreover, large-scale BECCS deployment requires extensive infrastructure for biomass management, carbon capture, and storage, which can be technically complex, costly, and time-consuming to develop and maintain. Policy and regulatory uncertainties, along with carbon pricing mechanisms, may also impede BECCS investment and progress (Gough et al., 2018). Additionally, public perception and social acceptance play crucial roles in BECCS implementation, as concerns about land use, biodiversity, food security, and environmental impacts may influence public attitudes towards this technology as a viable solution for carbon dioxide removal (Fridahl & Lehtveer, 2018).

4 Ocean Alkalinity Enhancement (OAE)

i. What is it?

OAE, also known as Ocean Fertilization (OF), involves the introduction of nutrients into the ocean to stimulate phytoplankton growth. Phytoplankton absorb CO2 during photosynthesis, leading to increased carbon uptake by the ocean. This process can be achieved by directly or indirectly increasing nutrient levels in surface waters, thereby enhancing the growth of microscopic marine plants (Williamson et al., 2012).

ii. Policy integration and implementation examples.

An example coming from Canada has investigated the possible consequences of ocean fertilization through iron sulphate. The experiment took place in 2021, when a fishing boat strewed 100 tonnes of iron sulphate into the ocean off western Canada (Tollefson, 2012). In 2021, the Haida Salmon Restoration Corporation (HSRC) started an ocean fertilization project. The HSRC tipped 120 tonnes of iron sulphate and iron oxide into an ocean eddy in international waters off the west coast of Haida Gwaii. The costs of this method are highly uncertain, and estimates vary from $70 and $160 per ton (Eisaman et al., 2023).

iii. Strengths/Opportunities.

The positive impacts of OAE are currently outweighed by its drawbacks. Through ocean fertilization it is possible to increase the number of elements, such as nitrogen, phosphorus, and iron, in the ocean. Those elements are nutrients for phytoplankton, which form the basis of the marine food web and have the ability to increase carbon capture (OceanNETs, 2024).

iv. Weaknesses/Threats.

According to studies conducted on OAE, this manipulation could create deoxygenated "dead zones" and deplete nutrients that would fuel phytoplankton growth in other areas (Geoengineering Monitor, 2021).

Some of the possible negative outcomes of this CDR technique can be exemplified by the case study offered by Haida Salmon Restoration Corporation. The idea behind the HSRC was to boost local salmon stocks while generating carbon credits (Gannon & Hulme, 2018). The experiment was not successful, and the release of iron sulphate ended up having a negative impact on the environment. The lack of scientific research and legal regulations are also considered causes of failure (Gannon & Hulme, 2018).

Another weakness of OAE is the lack of international legal support. Attempts to regulate ocean fertilization (OF) can be found in the maritime sector, with regard to ocean protection. The 1975 London Convention on the Prevention of Marine Pollution by Dumping of Wastes and Other Matter represents one of the first attempts to protect the maritime environment from human activities (International Maritime Organization, n.d.). This Convention was implemented in 2006 by the London Protocol, which, among other things, regulates climate change mitigation technologies such as marine geoengineering, specifically ocean fertilization activities (International Maritime Organization, n.d.). In the same year, an amendment to the London Protocol provided the basis in international environmental law to allow for carbon capture and storage beneath the seabed when it is safe to do so (Carbon Gap, n.d.). This is seen by the IPCC as a short-term technological option for reducing net CO2 emissions to the atmosphere (Carbon Gap, n.d.). A related amendment adopted in 2009 enables CO2 streams to be exported for CCS purposes, which can be provisionally applied following a decision in 2019 (International Maritime Organization, 2019). Moreover, in 2013, another amendment to the London Protocol was made to regulate marine geoengineering. Resolution LP.4(8) was adopted to allow ocean fertilization activities for research purposes only and to enable other marine geoengineering activities to be regulated in the future (National Oceanic and Athmospheric Administration, 2013).

5 Carbon Capture Utilization—Carbon Mineralization

i. What is it?

Carbon mineralization, a form of Carbon Capture Utilization (CCU), involves the reaction of carbon dioxide with different kinds of minerals to form solid carbonates, similar to chalk, antacids, or baking soda. This process converts CO2 into stable carbonates through chemical reactions with minerals. Through this process, carbon dioxide becomes bound in rocks as a solid mineral (World Resources Institute, 2023). While natural carbon mineralization occurs over long periods, research aims to develop accelerated mineralization processes to rapidly remove CO2 (National Energy Technology Laboratory, n.d.). Mineralization methods can be categorized as in-situ, occurring underground, or ex-situ, taking place above ground (National Academies of Sciences, Engineering, & Medicines, 2019).

ii. Policy integration and implementation examples.

An ongoing carbon mineralization project is being carried out in Australia by Mineral Carbonation International (MCI), a company founded in 2013 and headquartered in Canberra, Australia. MCI is owned by the GreenMag Group Pty Ltd., Orica Limited, and the University of Newcastle. MCI is building Carbon Plants that enable CO2 emissions reduction and removal, contributing to the fight against climate change. The mission of the company is to permanently remove a billion tonnes of CO2 by 2040 (MCi Carbon, 2024a). In 2017, a pilot plant facility was built in Newcastle, at the Newcastle Institute for Energy Research (NIER). The project was officially approved in April 2023 by Newcastle City Council (MCi Carbon, 2024b), and is one of the first mineral carbonation demonstration plants to convert thousands of tonnes of CO2 emissions annually into building products and other valuable materials for the circular economy. According to Ostovari et al. (2023), the assessments of the costs per ton of captured CO2 on plant level vary from $50 to $425.50.

iii. Strengths/Opportunities.

Manmade activities involving carbon mineralization techniques demonstrate the technical feasibility of carbon capture and mineralization processes. As previously mentioned, carbon mineralization can occur in two ways, in-situ and ex-situ. The first process consists of CO2 capture from power plants or industrial sources, which is then compressed and stored in reactive geologic formations where CO2 in the fluid form mineralizes to produce water-insoluble calcium or magnesium carbonates (Gadikota, 2021). The second process, ex-situ carbon mineralization, occurs when the captured CO2 is mineralized into carbonates in engineered processes. Ex-situ carbon mineralization can be developed as a strategy to capture and remove CO2 from energy and resource conversion processes that utilize fossil fuels and contribute to anthropogenic greenhouse gas emissions (Gadikota, 2021). The main positive impact of carbon mineralization comes from the reduction of greenhouse gas emissions from the atmosphere, but this is not the only effect. From an environmental point of view, carbon mineralization can bring to the formation of minerals that could improve soil quality and nutrient availability, contributing to agricultural productivity (Roper et al., 2012).

iv. Weaknesses/Threats.

Nevertheless, carbon mineralization has some drawbacks. Energy-intensive processes and reliance on certain energy sources may result in associated carbon emissions or environmental impacts (Cuéllar-Franca & Azapagic, 2015). Infrastructure development and operational costs can be significant (Styring et al., 2011). There could be considerable environmental trade-offs; for instance, extracting large quantities of minerals from natural deposits to support mineralization efforts can have ecological impacts, such as habitat disruption or depletion of finite mineral resources (Riedl et al., 2023; Styring et al., 2011).

6 SWOT Analysis for CDR Technologies

By examining the strengths, weaknesses, opportunities, and threats (SWOT) of CDR technologies as summarized in Fig. 2, we gain valuable insights into their potential implications and effectiveness in addressing climate change. Through this analysis, we seek to highlight both the promising aspects, and the challenges associated with the implementation of CDR strategies in mitigating greenhouse gas emissions and combating global warming.

6.1 Strengths

Natural Carbon Dioxide Removal (CDR) approaches can often offer relatively low costs as they can leverage existing infrastructure and resources, making them more accessible for implementation and scaling. For example, afforestation and reforestation projects utilize natural processes to remove carbon dioxide from the atmosphere, requiring lower upfront costs compared to technological solutions (Scientific American, 2021). In addition, blue carbon ecosystems have potential for natural carbon sinks, since mangroves, seagrasses, and saltmarshes have high carbon sequestration potential due to their ability to capture and store carbon in coastal environments (NOAA, n.d.). Similarly, enhancing soil health and adopting sustainable land management practices, such as cover cropping and conservation agriculture, contribute to carbon sequestration in terrestrial ecosystems (Hannam, 2024).

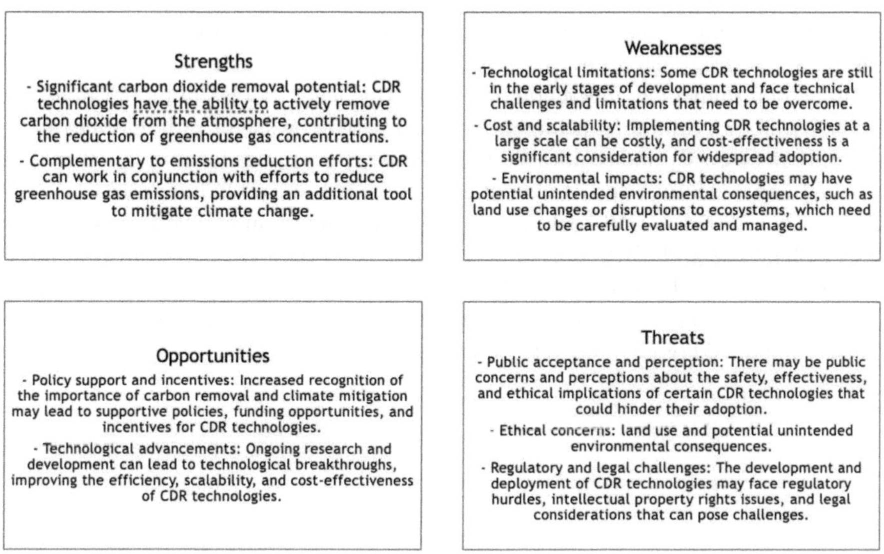

Fig. 2 A SWOT for CDR

Moreover, CDR technologies have the ability to actively remove carbon dioxide from the atmosphere and contribute to greenhouse gas reduction concentrations (IPCC, 2023). Versatility of the technologies enables tailored solutions that can address local carbon dioxide removal needs. CDR methods such as DACCS plants, which use chemical processes to capture carbon dioxide directly from the air, can be designed and implemented at different scales, from small pilot projects to large-scale installations (Küng et al., 2023).

Furthermore, CDR technologies can complement emissions reductions to accelerate near-term climate mitigations (World Business Council for Sustainable Development, 2023). CDR approaches also often offer additional co-benefits beyond carbon dioxide removal as they can enhance ecosystem services, improve soil health, promote biodiversity, and provide socio-economic advantages, making them attractive options for sustainable development. In this context, agroforestry practices for carbon sequestration not only capture carbon dioxide but also enhance soil fertility and water retention and provide opportunities for diversified livelihoods (Ghale et al., 2022). Therefore, the versatility and the flexibility of ground-based CDR offers the opportunity for integration with land management practices to enhance their effectiveness (Zoebisch et al., 2024).

6.2 Weaknesses

There are technical limitations and challenges of CDR approaches including scalability, efficiency, and reliability, which may affect their deployment on a large scale (CarbonLock, 2024). In addition, CDR technologies often require significant time to achieve meaningful carbon dioxide removal. Processes like afforestation and soil carbon sequestration require years or even decades to reach their full potential, limiting their immediate impact on carbon dioxide levels (Convention on Biological Diversity, 2012). Similarly, achieving long-term and consistent carbon removal and storage can be costly and require significant energy resources leading to lower CDR removal and concerns over its overall environmental impact (Chiquier et al., 2023; CarbonLock, 2024). CDR technologies may also have unintended consequences such as land use and disruption of ecosystems (CarbonLock, 2024). In fact, large-scale deployment of certain methods, such as afforestation or blue carbon projects, may require extensive land areas, which can be challenging to obtain, particularly in densely populated or developed regions (CarbonLock, 2024).

6.3 Opportunities

CDR technologies can also offer opportunities for local communities to participate and benefit from carbon dioxide removal efforts. Involving communities in the planning, implementation, and maintenance of projects can foster local support, knowledge sharing, and capacity building (Abouelnaga, 2022). Since CDR technologies are still in the early stages of development and thus require further research and refinement to improve their efficiency, scalability, and cost-effectiveness, there are many opportunities for breakthrough technological advancements (Mission Innovation, 2022).

Additionally, climate mitigation narratives and the crucial role of the CDR within them can contribute to public discussion and raise policy support, encouraging the development of the regulatory frameworks and their adoption. Likewise, government policies, carbon markets, and financial mechanisms can create favorable conditions for investment and implementation. (Honegger et al., 2021, Abouelnaga, 2022).

CDR also allows for synergies with Sustainable Development Goals (SDGs), such as coastal protection, biodiversity conservation in the case of maritime CDR, or sustainable agriculture and food security in the case of the land CDR (Honegger et al., 2021).

6.4 Threats

Another uncertainty is the environmental impact of climate change, meaning that sea level rise and coastal erosion can threaten the long-term viability and effectiveness of coastal CDR projects (NOAA Science Council, 2024). At the same time, soil degradation, erosion, and unsustainable land practices can reduce the carbon sequestration potential of soils, comprising the success of ground-based CDR efforts (Janssens et al., 2022).

Moreover, ground-based CDR technologies may have unintended environmental consequences or trade-offs and therefore ethical concerns. For instance, large-scale BECC projects could lead to land-use conflicts, competition for resources, or unintended impacts on ecosystems if not carefully managed (Congressional Research Service, 2024). A lack of awareness and buy-in from the public can also impede the successful implementation of CDR (Aziz et al., 2024). Finally, development and deployment of CDR without a proper regulatory framework can cause regulatory hurdles, intellectual property disputes, and other legal implications (Carbon Gap, 2024).

7 Conclusion

In conclusion, the SWOT analysis conducted for Carbon Dioxide Removal (CDR) technologies provides valuable insights into their strengths, weaknesses, opportunities, and threats in the context of addressing climate change.

The strengths of CDR technologies, such as relatively low costs, leveraging existing infrastructure, and offering a wide range of approaches, underscore their potential effectiveness in carbon dioxide removal. Additionally, their ability to provide additional co-benefits beyond carbon dioxide removal, such as enhancing ecosystem services and promoting socio-economic development, further enhances their appeal as sustainable solutions.

However, the SWOT analysis also highlights several weaknesses and threats associated with CDR technologies. Challenges such as spatial limitations due to land availability and suitability, potential environmental impacts, energy-intensive processes, and uncertainty in policy frameworks pose significant hurdles to their widespread adoption and implementation.

Despite these challenges, there are abundant opportunities for advancing CDR technologies. Engaging local communities, fostering policy support, and integrating CDR initiatives with Sustainable Development Goals (SDGs) can create favorable conditions for investment and implementation. Moreover, ongoing research and technological advancements offer prospects for overcoming existing limitations and enhancing the efficacy of CDR technologies.

In light of these findings, it is evident that a comprehensive approach is needed to harness the potential of CDR technologies while addressing their inherent challenges. By leveraging strengths, seizing opportunities, and mitigating weaknesses and threats, stakeholders can collectively work towards maximizing the impact of CDR technologies in mitigating climate change and building a sustainable future.

References

Abouelnaga, A. (2022). Community engagement in carbon dioxide removal initiatives. *Environmental Policy Review, 12*(3), 45–62.

Al-Juaied, M., & Whitmore, A. (2023). *Prospects for direct air carbon capture and storage: Costs, scale, and funding*. Belfer Center for Science and International Affairs, Harvard Kennedy School. Retrieved from December 15, 2024. https://www.belfercenter.org/publication/prospects-direct-air-carbon-capture-and-storage-costs-scale-and-funding

American University. (2020). *Fact sheet: Bioenergy with carbon capture and storage (BECCS)*. American University. Retrieved from December 15, 2024. https://www.american.edu/sis/centers/carbon-removal/fact-sheet-bioenergy-with-carbon-capture-and-storage-beccs.cfm

Aziz, M., Rahman, S., & Johnson, K. (2024). Public awareness and acceptance of carbon dioxide removal technologies. *Climate Policy Journal, 15*(2), 78–95.

Bui, M., Zhang, D., Fajardy, M., & Mac Dowell, N. (2021). Delivering carbon negative electricity, heat, and hydrogen with BECCS—Comparing the options. *International Journal of Hydrogen Energy, 46*(29), 15298–15321. https://doi.org/10.1016/j.ijhydene.2021.02.093

Canada Energy Regulator. (2022). *Market snapshot: New projects in Alberta could add significant carbon storage capacity by 2030*. Retrieved from December 15, 2024 https://www.cer-rec.gc.ca/en/data-analysis/energy-markets/market-snapshots/2022/market-snapshot-new-projects-alberta-could-add-significant-carbon-storage-capacity-2030.html
Canada Revenue Agency. (2024). *Carbon capture, utilization, and storage (CCUS) investment tax credit*. Retrieved from December 15, 2024. https://www.canada.ca/en/revenue-agency/services/tax/businesses/topics/corporations/business-tax-credits/clean-economy-itc/carbon-capture-itc.html
CanLII. (2010). *Carbon Capture and Storage Statutes Amendment Act, 2010*. Retrieved from December 15, 2024. https://www.canlii.org/en/ab/laws/astat/sa-2010-c-14/latest/sa-2010-c-14.html
Carbon Gap. (2024). Regulatory frameworks for carbon dioxide removal. *Legal Analysis Report, 8*(1), 123–145.
Carbon Gap. (n.d). *London Protocol*. Retrieved from December 15, 2024. https://tracker.carbongap.org/policy/london-protocol/
CarbonLock. (2024). *Human-Earth-System framework for assessing carbon dioxide removal strategies*. Retrieved from December 15, 2024. https://research.csiro.au/carbonlock/human-earth-system/
CCUS ZEN. (2024). *Exploring the potential for accelerating the deployment of CCUS in the Baltic Sea and Mediterranean Sea regions*. Retrieved from December 15, 2024. https://ccuszen.eu/
Chandler, D. (2019, October 25). *MIT engineers develop new way to remove carbon dioxide from air*. MIT News. Retrieved from December 15, 2024 https://news.mit.edu/2019/mit-engineers-develop-new-way-remove-carbon-dioxide-air-1025
Chiquier, S., Patrizio, P., Bui, M., Sunny, N., & Mac Dowell, N. (2023, December). *A comparative analysis of the efficiency, timing, and permanence of carbon dioxide removal (CDR) pathways*. AGU Fall Meeting Abstracts, 2023, GC54B-01.
ClimateSeed. (2024, January 30). *Carbon dioxide removals (CDR) methods*. Retrieved from December 15, 2024. https://climateseed.com/blog/carbon-removal-methods
Climeworks. (2021). *Orca: The world's first large-scale direct air capture and storage plant*. Retrieved from December 15, 2024. https://climeworks.com/plant-orca
Climeworks. (2022, June 28). *Mammoth: Our newest direct air capture and storage facility*. Retrieved from December 15, 2024. https://climeworks.com/plant-mammoth
Climeworks. (2023). *Supercharging carbon removal: A focus on direct air capture technology*. Climeworks Industry Snapshot. Retrieved from December 15, 2024. https://climeworks.com/uploads/documents/climeworks-industry-snapshot-3.pdf
Congressional Research Service. (2024). *Environmental impacts of large-scale carbon removal technologies*. CRS Report R47215. Washington, DC: Congressional Research Service.
Convention on Biological Diversity. (2012). *Impacts of afforestation on biodiversity*. Retrieved from December 15, 2024. https://www.cbd.int/doc/publications/cbd-ts-62-en.pdf
Cuéllar-Franca, R. M., & Azapagic, A. (2015). Carbon capture, storage, and utilization technologies: A critical analysis and comparison of their life cycle environmental impacts. *Journal of CO2 Utilization, 9*, 82–102. https://doi.org/10.1016/j.jcou.2014.12.002
Eisaman, M., Geilert, S., Renforth, P., Bastianini, L., Campbell, J., Dale, A., & Rønning, J. (2023). Assessing technical aspects of ocean alkalinity enhancement approaches. *State of the Planet Discussions, 2023*, 1–52. https://doi.org/10.21203/rs.3.rs-1765931/v1
Environmental, Social, and Governance Today. (2024, December 4). *Direct air capture provider Heirloom raises $150 million to scale carbon removal technology*. ESG Today. Retrieved from December 15, 2024. https://www.esgtoday.com/direct-air-capture-provider-heirloom-raises-150-million-to-scale-carbon-removal-technology/
European Commission. (2022a). *Commission staff working document executive summary of the impact assessment report:* Accompanying the document proposal for a regulation of the European Parliament and of the Council establishing a Union certification framework for carbon

removals (SWD (2022) 378 *final).* Retrieved from December 15, 2024. https://eur-lex.europa.eu/legal-content/EN/TXT/?uri=celex:52022SC0378

European Commission. (2022b). *Replication of innovative concepts for peri-urban, rural or inner-city mobility (RECIPROCITY).* CORDIS. Retrieved from December 15, 2024. https://cordis.europa.eu/project/id/101006656

European Commission. (2024). *Communication from the Commission to the European Parliament, the Council, the European Economic and Social Committee and the Committee of the Regions*: Towards an ambitious Industrial Carbon Management for the EU (COM/2024/62 final). Retrieved from December 15, 2024. https://eur-lex.europa.eu/legal-content/EN/TXT/?uri=CELEX:52024DC0062

Fajardy, M., & Greenfield, C. (2024). *Bioenergy with Carbon Capture and Storage-Energy System. IEA.* Retrieved from December 15, 2024. https://www.iea.org/energy-system/carbon-capture-utilisation-and-storage/bioenergy-with-carbon-capture-and-storage

Fajardy, M., & Mac Dowell, N. (2017). Can BECCS deliver sustainable and resource efficient negative emissions? *Energy & Environmental Science, 10*(6), 1389–1426.

Fajardy, M., Morris, J., Gurgel, A., Herzog, H., Mac Dowell, N., & Paltsev, S. (2020). *The economics of bioenergy with carbon capture and storage (BECCS) deployment in a 1.5°C or 2°C world* (Report No. 345). MIT Joint Program on the Science and Policy of Global Change. Retrieved from December 15, 2024. https://globalchange.mit.edu/sites/default/files/MITJPSPGC_Rpt345.pdf

Fern. (2022). *Six problems with BECCS.* Retrieved from December 15, 2024. https://www.fern.org/fileadmin/uploads/fern/Documents/2022/Six_problems_with_BECCS_-_2022.pdf

Fridahl, M., & Lehtveer, M. (2018). Bioenergy with carbon capture and storage (BECCS): Global potential, investment preferences, and deployment barriers. *Energy Research & Social Science, 42*, 155–165.

Gadikota, G. (2021). Carbon mineralization pathways for carbon capture, storage and utilization. *Communications Chemistry, 4*(1), 23.

Gannon, K. E., & Hulme, M. (2018). Geoengineering at the edge of the world: Exploring perceptions of ocean fertilisation through the Haida Salmon restoration corporation. *Geo: Geography and Environment, 5*(1), e00054.

Geoengineering Monitor. (2021). Ocean fertilization. Retrieved from December 15, 2024. https://www.geoengineeringmonitor.org/wp-content/uploads/2021/04/ocean-fertilization.pdf

Gerrard, M. B., & Hester, T. (Eds.). (2018). *Climate engineering and the law: Regulation and liability for solar radiation management and carbon dioxide removal.* Cambridge University Press.

Ghale, B., Mitra, E., Sodhi, H. S., Verma, A. K., & Kumar, S. (2022). Carbon sequestration potential of agroforestry systems and its potential in climate change mitigation. *Water, Air, & Soil Pollution, 233*(7), 228.

Gough, C., Garcia-Freites, S., Jones, C., Mander, S., Moore, B., Pereira, C., & Welfle, A. (2018). Challenges to the use of BECCS as a keystone technology in pursuit of 1.5 °C. *Global Sustainability, 1*, e5.

GOV.UK. (2023, March 15). *Government provides a springboard to UK CCUS industry with £20 billion for early deployment.* GOV.UK. Retrieved from December 15, 2024. https://www.gov.uk/government/news/government-reignites-industrial-heartlands-10-days-out-from-the-international-investment-summit

Hannam, I. (2024). Sustainable soil management and soil carbon sequestration. In H. Ginzky et al. (Eds.), *International yearbook of soil Law and policy 2022* (pp. 3–33). Springer.

Honegger, M., Michaelowa, A., & Roy, J. (2021). Potential synergies and trade-offs between carbon dioxide removal and sustainable development goals. *Environmental Research Letters, 16*(7), 074001.

Hossain, T., Burli, P., Pina, J., Jones, D., Hartley, D., & Hess, R. (2023). *Deployment of BECCUS value chains in the United States: A case study of sequestering CO2 from ethanol production.*

IEA Bioenergy. Retrieved from December 15, 2024. https://www.ieabioenergy.com/wp-content/uploads/2023/03/BECCUS-1.0_US-Case-Study_final_update.pdf
Inevitable Policy Response (IPR). (2023). Financing Direct Air Carbon Capture and Storage (DACCS): Quantifying the investment opportunity. Inevitable Policy Response. Retrieved from December 15, 2024. https://ipr.transitionmonitor.com/cms/wp-content/uploads/2023/11/IPR_DACCS_Financing_final.pdf
Intergovernmental Panel on Climate Change (IPCC). (2019). Technological and economic potential of options to enhance, maintain and manage biological carbon reservoirs and geo-engineering. In *Climate Change and Land: An IPCC Special Report on Climate Change, Desertification, Land Degradation, Sustainable Land Management, Food Security, and Greenhouse Gas Fluxes in Terrestrial Ecosystems*. Retrieved from December 15, 2024. https://www.ipcc.ch/site/assets/uploads/2018/03/4.pdf
Intergovernmental Panel on Climate Change (IPCC). (2022). Carbon dioxide removal (CDR) fact-sheet. In *Climate Change 2022: Mitigation of Climate Change*. Retrieved from December 15, 2024. https://www.ipcc.ch/report/ar6/wg3/downloads/outreach/IPCC_AR6_WGIII_Factsheet_CDR.pdf
Intergovernmental Panel on Climate Change. (2023). IPCC AR6 WGIII: *CDR Factsheet Carbon Dioxide Removal*. Retrieved from December 15, 2024. https://www.ipcc.ch/report/ar6/wg3/downloads/outreach/IPCC_AR6_WGIII_Factsheet_CDR.pdf
International Energy Agency (IEA). (2022a). *Bioenergy with carbon capture and storage*. IEA. Retrieved from December 15, 2024. https://www.iea.org/energy-system/carbon-capture-utilisation-and-storage/bioenergy-with-carbon-capture-and-storage
International Energy Agency (IEA). (2022b). *Direct Air Capture-Energy System*. IEA. Retrieved from December 15, 2024. https://www.iea.org/energy-system/carbon-capture-utilisation-and-storage/direct-air-capture
International Energy Agency. (2022c). Inflation Reduction Act 2022: Sec. *13104 extension and modification of credit for carbon oxide sequestration*. International Energy Agency. Retrieved from December 15, 2024. https://www.iea.org/policies/16255-inflation-reduction-act-2022-sec-13104-extension-and-modification-of-credit-for-carbon-oxide-sequestration
International Energy Agency Greenhouse Gas R&D Programme [IEAGHG]. (2021). *Global Assessment of Direct Air Capture Costs*. Retrieved from December 15, 2024. https://ieaghg.org/publications/global-assessment-of-direct-air-capture-costs/
International Maritime Organization. (2019). 41st consultative meeting of contracting parties to the London convention and the 14th meeting of contracting parties to the London protocol.
International Maritime Organization. (n.d). *Convention on the Prevention of Marine Pollution by Dumping of Wastes and Other Matter*.
Janssens, I. A., Van Der Meer, M., & Schmidt, M. W. (2022). Soil degradation and carbon sequestration potential: Global challenges and regional responses. *Soil Biology and Biochemistry, 165*, 108502.
Japanese Ministry of Economy, Trade and Industry. (2023, June 13). *Full-scale commencement of Japanese CCS projects*. Ministry of Economy, Trade and Industry. Retrieved from December 15, 2024. https://www.meti.go.jp/english/press/2023/0613_001.html
Kriegler, E., Edenhofer, O., Reuster, L., Luderer, G., & Klein, D. (2013). Is atmospheric carbon dioxide removal a game changer for climate change mitigation? *Climatic Change, 118*(1), 45–57.
Küng, L., Aeschlimann, S., Charalambous, C., McIlwaine, F., Young, J., Shannon, N., & Garcia, S. (2023). A roadmap for achieving scalable, safe, and low-cost direct air carbon capture and storage. *Energy & Environmental Science, 16*(10), 4280–4304.
MCi Carbon. (2024a). *Our mission*. Retrieved from December 15, 2024. https://mcicarbon.com/
MCi Carbon. (2024b). *Technology*. Retrieved from December 15, 2024. https://mcicarbon.com/technology/
Mission Innovation. (2022). *Carbon dioxide removal innovation challenges and opportunities*. Mission Innovation Report. Retrieved from December 15, 2024. https://mission-innovation.net/wp-content/uploads/2022/05/CDR-Innovation-Report.pdf

MIT Climate Portal. (n.d). *Enhanced rock weathering.* MIT Climate Portal. Retrieved from December 15, 2024. https://climate.mit.edu/explainers/enhanced-rock-weathering

Mulabagal, V., Baah, D. A., Egiebor, N. O., & Chen, W. Y. (2022). Biochar from biomass: A strategy for carbon dioxide sequestration, soil amendment, power generation, and CO2 utilization. *Handbook of climate change mitigation and adaptation, 4*, 1937–1974.

National Academies of Sciences, Engineering, and Medicine (U.S.). Committee on Developing a Research Agenda for Carbon Dioxide Removal and Reliable Sequestration. (2018). *Bioenergy with carbon capture and sequestration in Negative emissions technologies and reliable sequestration: A research agenda.* The National Academies Press. Retrieved from December 15, 2024. https://www.ncbi.nlm.nih.gov/books/NBK541444/

National Academies of Sciences, Engineering, and Medicine. (2019). *Negative emissions technologies and reliable sequestration: A research agenda.* The National Academies Press.

National Energy Technology Laboratory. (n.d). *Carbon mineralization.* Retrieved from December 15, 2024. https://netl.doe.gov/carbon-management/carbon-storage/mineralization

National Oceanic and Atmospheric Administration (NOAA). (2013). *Resolution LP. 4(8) on the Amendment to the London Protocol to Regulate the Placement of Matter for Ocean Fertilization and Other Marine Geoengineering Activities.* Retrieved from December 15, 2024. https://www.gc.noaa.gov/documents/resolution_lp_48.pdf

Natural Resources Canada. (2023). Canada's Carbon Management Strategy. Retrieved from December 15, 2024. https://natural-resources.canada.ca/climate-change/canadas-green-future/capturing-the-opportunity-carbon-management-strategy-for-canada/canadas-carbon-management-strategy/25337

Nayak, N., Mehrotra, R., & Mehrotra, S. (n.d). Carbon biosequestration strategies: A review. *Carbon Capture Science & Technology, 4*, 100065.

NOAA Science Council. (2024). Climate change impacts on coastal ecosystems: Sea level rise and coastal erosion. NOAA Technical Report NMFS-456. Seattle, WA: NOAA.

NOAA. (n.d). What is blue carbon? Retrieved from December 15, 2024. https://oceanservice.noaa.gov/facts/bluecarbon.html

OceanNETs. (2024). Ocean fertilization and its potential for carbon capture. Retrieved from December 15, 2024. https://www.oceannets.eu/ocean-fertilization

Ostovari, H., Sternberg, A., & Bardow, A. (2023). Techno-economic assessment of carbon capture and utilization pathways. *Journal of Cleaner Production, 234*, 117–129.

PNP Charcoal. (2024). *How is charcoal made*? PNP Charcoal website. (Retrieved from December 15, 2024).

Riedl, D., Z. Byrum, S. Li, H. Pilorgé, P. Psarras, & K. Lebling. (2023). *5 Things to Know About Carbon Mineralization as a Carbon Removal Strategy.* Retrieved from December 15, 2024. https://www.wri.org/insights/carbon-mineralization-carbon-removal

Rodrigues, C. I., Dias, L. M., & Nunes, L. J. (2023). Soil carbon sequestration in the context of climate change mitigation: A review. *Soil Systems, 7*(3), 64. https://doi.org/10.3390/soilsystems7030064

Roper, M. M., Milroy, S. P., & Poole, M. L. (2012). Green and brown manures in dryland wheat production systems in Mediterranean-type environments. *Advances in Agronomy, 117*, 275–313.

Scientific American. (2021). *The real cost of planting trees.* Retrieved from December 15, 2024. https://www.scientificamerican.com/article/the-real-cost-of-planting-trees/

Scottish Carbon Capture & Storage. (n.d). About us. Retrieved from December 15, 2024. https://www.sccs.org.uk/about-us

Styring, P., Jansen, D., De Coninick, H., Reith, H., & Armstrong, K. (2011). *Carbon capture and utilisation in the green economy* (p. 60). Centre for Low Carbon Futures.

Thangaraj, P., Okoye, S., Gordon, B., Zilberman, D., & Hochman, G. (n.d). *Bioenergy with carbon capture and storage (BECCS): Reaching global goals of energy security and climate change abatement.* USDA website. (Retrieved from December 15, 2024).

Tollefson, J. (2012). Ocean-fertilization project off Canada sparks furore. *Nature, 490*(7421), 458.

U.S. Department of Energy. (2022). *The Inflation Reduction Act drives significant emissions reductions and positions America to reach our climate goals.* U.S. Department of Energy. Retrieved from December 15, 2024. https://www.energy.gov/sites/default/files/2022-08/8.18%20InflationReductionAct_Factsheet_Final.pdf

University of Edinburgh. (2024). *HyStorPor: Hydrogen Storage in Porous Media.* Retrieved from December 15, 2024. https://blogs.ed.ac.uk/hystorpor/

Vandermel, J. (2020). Preventing Climate Change with BECCS: Bioenergy with Carbon Capture and Storage. In *Princeton Student Climate Initiatives*; Princeton University.

Wesche, J. P., & Skjølsvold, T. M. (2024). Direct air capture technologies: Location independence and technological innovation potential. *Energy Policy, 186*, 113995.

Williamson, P., Wallace, D. W. R., Law, C. S., Boyd, P. W., Collos, Y., Croot, P., et al. (2012). Ocean fertilization for geoengineering: A review of effectiveness, environmental impacts and emerging governance. *Process Safety and Environmental Protection, 90*(6), 475–488. https://doi.org/10.1016/j.psep.2012.10.007

World Bank. (2023). *Blue carbon ecosystems: Mangroves, seagrasses, and salt marshes.* World Bank Technical Report. World Bank.

World Business Council for Sustainable Development. (2023). *Carbon dioxide removal technologies: Complementing emission reductions for climate mitigation.* WBCSD Report. Geneva: WBCSD.

World Economic Forum. (2023). *Direct air capture technology: Chemical processes for carbon removal.* World Economic Forum White Paper. WEF.

World Resources Institute. (2023). *Carbon mineralization: Converting CO2 into stable carbonates.* Retrieved from December 15, 2024. https://www.wri.org/insights/carbon-mineralization-carbon-removal

World Resources Institute. (2024). *Direct air capture technology: Land footprint analysis and comparison with natural methods.* WRI Technical Report. WRI.

Zoebisch, M. A., Chen, K. L., & Rahman, F. (2024). Integration of carbon dioxide removal with sustainable land management practices. *Land Use Policy, 138*, 107045.

SWOT Analysis for Solar Radiation Modification (SRM)

Valeria Pinna, Leonora Burnik, and Laure-Marine Vioujard

Abstract Solar Radiation Modification (SRM), also known as solar geoengineering, is a geoengineering method that entails any intentional efforts to increase the amount of sunlight reflected in space, thereby reducing the amount of sunlight absorbed by the Earth. This form of climate intervention seeks to counteract global warming by enhancing the Earth's ability to reflect sunlight into space before it contributes to heat absorption in the environment. This chapter offers an examination of some of the SRM methods that have been investigated by the scientific community to attenuate climate change. These methods are divided into two categories: stratospheric and space-based, and ground-based technologies. To perform the analysis, it was chosen to follow the SWOT analysis approach.

Keywords Solar Radiation Modification · Solar Geoengineering · Climate Change · SWOT analysis

1 What Is SRM, Overall Analysis and Elements

Solar Radiation Modification (SRM) refers to a set of proposed technologies and techniques aimed at reflecting or blocking a portion of the sun's energy to cool the Earth's surface and counteract the effects of climate change. Stratospheric- and space-Based SRM and Ground-Based SRM are two categories of techniques within the broader field of SRM (European Commission's Group of Chief Scientific Advisor, 2013). This chapter begins by providing an explanation of each category of SRM, giving the definitions, elements, and general overview of the methods. Subsequently, each method will be examined in greater detail, defining what it is, giving a short

V. Pinna (✉)
European Union Agency for the Space Programme, Prague, Czech Republic

L. Burnik
European Union Agency for the Space Programme, Prague, Czech Republic

L.-M. Vioujard
European Union Agency for the Space Programme, Prague, Czech Republic

K.-U. Schrogl et al. (eds.), *The Politics of Geoengineering*, Springer Climate,
https://doi.org/10.1007/978-3-032-10095-5_3

overview of existing policies (if applicable), outlining the existing plans, and giving examples of the already implemented studies. Moreover, each method will be analysed according to the SWOT analysis. Finally, the chapter will conclude with a comprehensive SWOT analysis of the SRM, summarizing the potential advantages and disadvantages inherent in this approach.

1.1 Stratospheric- and Space-Based SRM

Stratospheric- and space-based SRM refers to the use of technologies deployed in space and the upper layer of the atmosphere to modify the Earth's energy balance and reduce the amount of solar radiation reaching the planet's surface. Some proposed concepts include:

A. Orbital Sunshades: This approach involves placing large reflective structures or shades in space to block or redirect a portion of the incoming solar radiation. These shades would act as a barrier between the sun and the Earth, reducing the amount of sunlight reaching the planet's surface (Fuglesang & Miciano, 2021).
B. Stratospheric Aerosol Injection (SAI): SAI involves injecting reflective aerosols, such as sulphur dioxide (SO2), into the stratosphere from the Earth's surface. The aerosols scatter sunlight back into space, thereby reducing the amount of solar radiation reaching the surface (Acker et al., 2022).

1.2 Ground-Based SRM

Ground-based SRM techniques focus on modifying the Earth's environment at the surface or within the lower atmosphere to reduce the amount of solar radiation absorbed by the planet. Some proposed ground-based SRM methods include:

A. Land Surface Modifications: This is a local-scale ground-based SRM which entails modifying the Earth's surface to increase its reflectivity and reduce the absorption of solar radiation. This can be achieved through various methods, such as using reflective materials on buildings and surfaces, implementing light-coloured pavements or roofs, or adopting land modification practices that enhance surface albedo (reflectivity) (Santamouris, 2014).
B. Marine Cloud Brightening: This technique aims to increase the reflectivity of marine clouds by injecting seawater aerosols or other cloud condensation nuclei into the lower atmosphere. The enhanced cloud reflectivity would result in more sunlight being reflected into space (Feingold et al., 2024).

Both techniques are still largely theoretical and face significant technical, ethical, and governance challenges. Extensive research and assessment are necessary to understand their significant uncertainties and potential impacts and risks before any practical implementation could be considered. Furthermore, as for all the geoengineering techniques, SRM technologies are intended to be complemfigentary to, rather than a substitute for, mitigation efforts aimed at reducing greenhouse gas emissions.

2 SRM Single Techniques—SWOT Analysis

This paragraph will showcase some stratospheric- and space-based techniques that are under investigation by scientists and private companies around the world. However, before going into details, it is also important to give an overview of recent developments in international discussions on SRM. In recent years there has been an increase in public visibility of and discussions on SRM technologies in general and their governance and regulation in particular.

In 2022, the World Meteorological Organization, a specialized agency of the UN published *Scientific Assessment of Ozone Depletion* in which they recognized SAI's potential impact of reducing global temperatures but also warned of potential risks and identified gaps in knowledge and uncertainties to give a complete evaluation (World Meteorological Organization, 2022). Moreover, in June 2023, the White House published a *Congressionally Mandated Research Plan and an Initial Research Governance Framework Related to Solar Radiation Modification* focusing on the research needed to improve understanding of the potential impacts of SRM. The research plan also addresses the methodologies for conducting the research, including the roles of Federal initiatives and international cooperation (The White House, 2023).

Similarly, in June 2023 a Joint Communication to the European Parliament and the Council, titled "*A new outlook on the climate and security nexus: Addressing the impact of climate change and environmental degradation on peace, security and defence*" was published. It acknowledged that SRM technologies are increasingly becoming more prominent, and warned of the possible risks, impacts, and unintended consequences these technologies could have. Moreover, it called for precaution and discussion on the international framework on research and governance of SRM (European Commission, 2023).

Furthermore, in February 2023, a group of more than 110 physical and biological scientists wrote a *Letter of support for research on atmospheric aerosols and their potential to increase the reflection of sunlight from the atmosphere to address climate risk,* in which they also called for more research on SRM, its risks, impacts and information relevant for suitable governance (Members of the Physical and Biological Community, 2023). On the contrary, in January 2022 a group of more than 450 scientists signed an open letter from an initiative for *International Non-Use Agreement on Solar Geoengineering.* The initiative is directed at governments, the UN, and other stakeholders to prevent the normalization of SRM as a climate policy option (International Non-Use Agreement on Solar Geoengineering, 2022).

In September 2023, the Climate Overshoot Commission published a report *Reducing the Risks of Climate Overshoot* in which it called upon, among others, the international community to, on the one hand, adopt a moratorium on the deployment of SRM technologies and large-scale field experiments which would carry risk of significant transboundary harm; on the other, to further expand research on technologies and open international governance dialogues (Climate Overshoot Commission, 2023). Finally, after the unauthorized SAI outdoor experiment conducted by the US start up Making Sunsets in Mexico in 2023, their government announced to develop and implement a legal framework to prevent any solar geoengineering experiments (Rodriguez & Lo, 2023).

2.1 *Stratospheric- and Space-Based Techniques*

Orbital Sunshades

(i) What is it?

Orbital sunshades are all geoengineering techniques that involve the placement of a large structure or a constellation of reflecting objects in space to block incoming solar radiations or reflect sunlight away from the Earth. Those objects would be placed in space through the launch of rockets and positioned at a Lagrange point between the Sun and the Earth (Boyd, 2024). The purpose of orbital sunshades is to mitigate the effects of global warming and the consequences of climate change by reducing the amount of solar radiation reaching the Earth's surface. Since the twentieth century, several proposals have been studied to make use of solar sunshades to cool the Earth and even to positively influence countries' war strategies. In 1960, for example, Soviet scientists proposed the placement of rings of dust in the upper atmosphere to reflect sunlight onto the planet to warm certain areas, a proposal that was in line with the attempts to warm Siberia by rerouting Russian rivers (Kintisch, 2018).

Orbital sunshades could be implemented through a variety of methods, such as:

- Sunshade Satellites: It consists of satellites or constellations of satellites equipped with reflective materials to reduce Solar radiation. Studies on sunshade satellites date back to the 20 s when physician Hermann Oberth started to investigate this method to manipulate Earth's weather.
- Space-based Mirrors: It is based on the installation of large mirrors or reflective panels in space to reduce the amount of solar energy reaching the Earth.
- Aerosols or Dust Particles: It involves the spreading of fine particles in a stable orbit around the Earth to scatter sunlight away from the planet and reduce its absorption (Irvine et al., 2016).

(ii) Policy integration and implementation examples.

A concrete example of how to implement SRM climate engineering techniques is currently being carried out by the Massachusetts Institute of Technology (MIT) scientists concerning the so-called "space bubbles' '(MIT, 2022). The team of researchers believes that space bubbles could be one of the most efficient thin-film structures for deflecting solar radiation. According to their research, these structures could be manufactured directly in outer space and placed strategically at the Lagrangian Point between the Earth and the Sun. The thin film of silicon bubbles deployed into outer space would roughly have the size of Brazil and would contribute to blocking the Sun's solar radiation from further warming Earth and reverse climate change (Newcomb, 2022). The estimated costs of these SRM approaches are considerable and various numbers have been reported from $1 trillion, a few trillion USD or from $6 to 20 trillion for deployment of an initial sunshade system (Baum et al., 2022; Fuglesang & de Herreros Miciano, 2021).

(iii) Strengths/Opportunities.

As for all the climate engineering methods, orbital sunshades have the goal to mitigate the consequences of climate change and manipulate nature. In this case, as a solar radiation modification technique, solar sunshades try to reduce the direct impact of sunshine and Solar radiation on the Earth's environment, consequently reducing global warming. The positive impact of those techniques consists of a significant reduction in the Earth's average temperature (Coverstone & Johnson, 2015).

(iv) Weaknesses/Threats.

When addressing climate engineering, weaknesses and threats should be taken into account and, in particular, the significant technical, ethical, and environmental challenges associated with its implementation. The deployment of orbital structures and mirrors in space is foreseen to have high costs. Reflective objects should be placed on the L1 Lagrange point, at about 1.5 million km from the Earth, with an estimated cost of launch of hundreds of billions of dollars (Siegel, 2018). An estimation of costs comes from the University of Arizona where, in 2006, one of the first orbital sunshade projects was investigated by the astronomer and regent's Professor Roger Angel. According to Prof. Angel's analysis, the feasibility of a space sunshade project would take about 25 years of work for an investment of a few trillion dollars (Stiles, 2006). Therefore, geoengineering solutions need enormous financial investment, without considering the elevated risks connected to them. The potential consequences on Earth's ecosystems are many, even if not yet fully understood. For instance, mirrors and sunshade objects could lead to cooler tropics and warmer poles and eventually reduce the gradient between the tropics and poles and affect the seasonal cycle (Kintisch, 2018).

Stratospheric Aerosol Injection (SAI)

(i) What is it?

Stratospheric Aerosol Injection (SAI) or Stratospheric Aerosol Albedo Modification (SAAM) is the most proposed SRM method that offers the possibility to offset some of the consequences of global warming at an affordable cost (Kintisch, 2018). However, SAI's goal is not to address the underlying causes of climate change but to control the amount of incoming solar radiation (Paddison, 2023). The process of achieving this includes injecting inorganic particles, such as calcium carbonate or sulphur dioxide, into the stratosphere to create a protective, reflective barrier, that will ultimately lead to a reduction in the quantity of solar radiation reaching the Earth's surface and decrease the average air surface temperature (Surprise, 2019). It was first proposed in 1974 by Russian scientist Mikhail Budyko, and then in 2006 Nobel laureate Paul Crutzen gave it legitimacy by advocating for further research (Kintisch, 2018).

The concept of SAI is based on replicating the cooling effect of the volcanoes, which have historically proven to be effective in decreasing the temperature on the Earth's surface. Such consequences have been detected after the eruption of Mount Pinatubo in the Philippines in June of 1991. The release of 20 million tons of sulphur dioxide into the stratosphere elevated the Earth's albedo, which led to a subsequent decrease in the amount of absorbed solar radiation and resulted in estimated cooling of average air surface temperature of 0.3° over three years. However, although the cooling down occurs fast, usually within a year following the change of reflectivity, it is not permanent, as it lasts only for a few years unless the particle injection process is perpetual.

To achieve the same or similar results, various proposals have been put forth, from designing self-levitating particles, to using titanium, aluminium, calcite, or diamond dust and launching them via artillery guns, large hoses or from the rear of aircraft. The most studied and tested option so far is sulphates-based SAI, however, like all others, is still in the research phase with recent changes. Furthermore, modelling studies suggest that to counterbalance the climate effects of the CO2 the SAI approach would need to block approximately 1,7 to 2,5% of solar radiation. One of the estimated yearly costs over the first 15 years of a deployment of SAI using an aircraft-based delivery system is an average cost of ~$2.25 billion yr.—1, while others suggest that its costs would be about $18 billion yr.—1 per degree Celsius of warming avoided (Smith & Wagner, 2018; Smith, 2020).

(ii) Policy integration and implementation examples.

The Stratospheric Welsbach Seeding method was developed and patented in 1991 by Hughes Aircraft Company, a subsidiary of General Motors and Raytheon, and it proposed seeding the stratosphere with small oxide particles (e.g., thorium, barium, or aluminium flakes) to reflect sunlight and reduce global warming. However, it is not known publicly if the implementation of the proposal was put forward (Chang & Shih I-Fu., 1991). Another research project, "Collaborative Research: Evaluation of Suggestions to Geoengineer the Climate System Using Stratospheric Aerosols and Sun Shading", has been developed by the School of Environmental and Biological Sciences at the University of California in Los Angeles. The project went underway between 2008 and 2012, and it modelled how shading caused by stratospheric aerosol injections would affect the climate and analysed its political and ethical considerations. It was funded by the U.S. National Science Foundation, and the reported awarder funding was $77,571.00 (U.S. National Science Foundation, 2009).

More research on SAI has been conducted outside the USA as well. In Russia, a series of field tests has been conducted by a group of scientists from the Institute of

Global Climate and Ecology (Russian Academy of Sciences), the state Hydrometeorological Service, the Central Aerological Observatory, and the industrial association "Taifun" (Oldfield & Poberezhskaya, 2023). In the experiment, sulphate particles were released both from a truck and from a helicopter provided by the Russian military, and then the solar radiation in the visible wavelength was measured (Izrael et al., 2009).

Another project was the EU-funded initiative "Implications and risks of engineering solar radiation to limit climate change (IMPLICC)", that was running between 2009 and 2012 and was studying the efficiency, side effects, and economic implications of the three main SRM technologies: space-based mirrors, sulphur injections into the stratosphere and low-level marine clouds. The project received funding of €1.32 million, mostly from the EU, and was coordinated by Max Planck Gesellschaft Zur Forderung Der Wissenschaften Ev (European Commission, 2013).

More recently, as abovementioned, the Making Sunsets. Inc. company launched a sulphur-based SAI experiment from two sites in the northern Mexican state of Baja California in January 2023 and later moved to the USA when the Mexican government banned unauthorized experiments from its territory. Consequent field tests happened in Reno in February 2023 and in San Francisco in April 2023. Their SAI method is based on the concept of balloons filled with helium and a small amount of SO2 that, once released in the air, would burst and inject the sulphur particles into the stratosphere and create reflective clouds. Furthermore, the company was selling the so-called cooling credits for balloon flights containing higher amounts of sulphur dioxide for 10$. Their experiments led to two outcomes: on the one hand, Mexico proposed to adopt new regulations and standards aimed at banning solar geoengineering, and on the other, due to the small number of released particles, there was no detected threat to the environment. The company is funded by the venture capital fund Boost VC (US$ 0.5 M, 2022), Draper Associates (since 2023), Pioneer Fund (US$ 0.2 M, 2022), family/friends (US$ ~ 0.5 M), angel investors (Geoengineering Monitor, 2021a).

Additionally, Harvard University has been conducting the Stratospheric Controlled Perturbation Experiment (SCoPEx) under its Harvard Solar Geoengineering Research Program (SGRP) (Sapinski et al., 2021). SCoPEx entailed researching, modelling, and eventual field testing of SAI by dispersing particles such as ice, calcium carbonate, sulphate and other materials from an instrument placed in a gondola. This gondola, attached to a high-altitude balloon, was controlled from the ground and equipped with data-collecting sensors (Geoengineering Monitor et al., 2021b). The aim was to predict the large-scale effects of SAI. Field testing was first envisioned in the USA, New Mexico, and Arizona, and then moved to Kiruna. However, in 2021, due to the strong opposition of the Saami Council, Swedish civil society and scientists, the field testing was cancelled. Generally, experimental hardware and operations for SCoPEx are funded by the internal research funds from Harvard University provided to Professors David Keith and Frank Keutsch, and additional research funding is provided by SGRP (Tollefson, 2021, March 31). In this case, the project also received approximately $16 million in private philanthropic donations from myriad foundations and private billionaires close to Silicon Valley and Wall Street, including Bill Gates who reportedly matches 40% of each gift (Keith, 2020).

(iii) Strengths/Opportunities.

SAI offers the potential to reduce global temperatures in a short time. It has instantaneous effects in the geological sense, making it serve as an emergency solution to address urgent climate change issues and give time for more comprehensive mitigation and adaptation strategies. Moreover, due to its structure, SAI opens the possibilities for flexible implementation, meaning that its quick response time allows for a rapid adjustment according to changing environmental conditions. Another advantage is its low economic costs, which make it potentially economically viable for global use regardless of the socio-economic context (Smith et al., 2018). Furthermore, more research on the SAI, particularly in regard to the use of aircraft used to inject and disperse the particles, could lead to the development of the aerospace industry. SAI could also have the capacity to influence the weather patterns positively, offering opportunities to solve drought issues and increase agricultural productivity in certain regions. Moreover, some studies showed that in comparison with no geoengineering, storm surge events comparable to ones caused by the 2005 Katrina hurricane could be reduced by about 50%, but this reduction is only marginally statistically significant. In addition to that, SAI projections suggest that it could reduce coastal flood risk by 40 cm for 5-y events and approximately half for 50-y surges (Moore et al., 2015).

(iv) Weaknesses/Threats.

On the other hand, SAI could lead to disrupted rain and wind patterns, altering precipitation and wind patterns across the tropics and subtropics (Patel, 2023). This could lead to unintended consequences such as droughts, disrupted monsoons, and other unplanned side effects. Moreover, the injection of aerosols could lead to unpredictable interactions with atmospheric dynamics and potentially exacerbate existing climate challenges. Such events could happen especially with the use of sulphate particles, which hold the potential to contribute to ozone layer depletion, resulting in impacts both on human health and the ecosystems. Consequently, this aerosol injection could introduce environmental and health risks by causing air pollution and acid rain. Furthermore, the use of SAI methods in one part of the world could cause unintended climate issues in other parts. For example, some studies suggest that SAI concentrated in the northern hemisphere would likely cause a reduction in rainfall over the Sahel area and vice versa for SAI in the southern hemisphere (Joneset al., 2017). Finally, some studies showed that SAI could cause deepening of the Antarctic ozone hole and delay in ozone recovery (World Meteorological Organization et al., 2022).

Apart from its impact on human health and ecosystems, SAI entails many ethical and political concerns. Firstly, the prospects of unilateral deployment of SAI, without international consensus or collaboration, could lead to geopolitical tensions and disputes (Tang & Kemp, 2021). Secondly, the development and implementation of SAI, even though it comes at a lower price compared to other geoengineering methods, is still considerable, and could lead to technical and resource challenges as it necessitates, for example, the utilization of advanced aircraft technology or acquisition of substantial quantities of aerosol materials. Moreover, by creating a reflective barrier, SAI might involve trade-offs with solar energy production by diminishing the amount of solar radiation that reached the Earth's surface and thus jeopardizing the transition towards renewable energy resources. Finally, there are uncertainties surrounding the commercial feasibility of SAI, as its potential for direct profitability is limited compared to other methods such as CDR (Smith et al., 2018).

2.2 *Ground-Based Techniques*

Land Surface Modifications

(i) What is it?

Land Surface Modification or Surface Albedo Modification is a solar geoengineering technique aimed at counteracting global warming by increasing the Earth's albedo or in other words increasing the amount of solar radiation reflected by a surface (Akbari, 2009). The idea behind this technique is to alter the reflectivity of various landscapes, such as ice, crops, forests, and deserts, to reflect more sunlight into space, thus reducing the amount of solar energy absorbed by the Earth's surface (Irvine et al., 2011). The goal of this technique is to temporarily mitigate rising temperatures and provide time for other climate mitigation strategies to take effect.

(ii) Policy integration and implementation examples.

Surface Albedo Modification has gained prominence as a potential tool within climate policies and strategies, with various experimental projects exploring diverse methods for altering surface reflectivity. For example, the Arctic Ice Project (formerly Ice911) seeks to insulate melting Arctic ice by applying reflective silica glass, albeit with raised concerns about ecological repercussions and material effects. The project has had approximately 250 donors, both private and from foundations, including NASA. Its budget has varied from $0.5 million in 2017, $0.8 million in 2018, $1.04 million in 2019, and $0.97 million in 2020. In 2021, it was reported that a multi-year and multi-million-dollar collaboration with Norwegian SINTEF was initiated (Arctic Ice Project, 2023).

Furthermore, the Swiss Academia Engiadina together with Dutch researcher Hans Oerlemans has tested the concept of deploying reflective artificial snow on glaciers to curb their melting, however, this approach demands substantial water and energy resources. They conducted a small pilot demonstration in 2017 where they covered 20mx20m of Diavolezzafirn glacier with 2.5 m of artificial snow. The cost of the study was $0.1 million. In the same year, they also conducted a small outdoor trial of covering Morteratsch Glacier in 2017 and received further funding for 30 months from August 2019 to January 2022 of 2,five million Swiss Franc, half of which is supported by Innouseisse (Fachhochschule Graubunden, 2019).

Moreover, Ohio State University's Greenland Proposal involves covering sections of the Greenland ice sheet with reflective materials to mitigate melting; however, this method entails the consumption of fossil raw materials and high costs. For instance, the costs for materials covering 1 km2 of glacier ice with reflective blankets was estimated at $ 4.6 million (Project Earth, 2008). Furthermore, the research project carried out by the University of Bristol proposes developing high-albedo crops with more reflective leaves, potentially aiding atmospheric cooling through increased solar radiation reflection, though implications for plant nutrition and soil remain uncertain (Geoengineering Monitor, 2021; Kala & Hirsch, 2022). Finally, the theory of Boreal Forest Clearing suggests that removing boreal forests in specific areas could have a cooling effect as snow-covered surfaces reflect more sunlight, but this raises worries about ecosystem disruption and climate regulation implications (DeFusco, 2011).

(iii) Strengths/Opportunities.

Surface Albedo Modification holds the potential to contribute to a temporary reduction in global temperatures by effectively reflecting more sunlight into space. The technological diversity within this approach offers adaptability and flexibility, enabling the implementation of localized interventions tailored to specific environmental contexts. Moreover, if well-managed, this technique could serve as a short-term climate mitigation strategy, providing a window of opportunity to transition towards cleaner energy sources and address the challenges of global warming.

(iv) Weaknesses/Threats.

Despite its potential, the alteration of surface albedo could lead to unintended ecological consequences, disrupt natural processes, and disturb established weather patterns, resulting in ecological imbalances that are difficult to predict (Irvine et al., 2016; National Research Council, 2015). The limited global impact of this approach, particularly if not implemented extensively, raises questions about its effectiveness in addressing broader climate challenges (Lawrence Mark, 2018). Furthermore, there are some ethical and societal concerns connected to this technique, ranging from the potential for ecosystem disruption, and resource conflicts, to causing adverse effects on communities dependent on these altered environments Keith & MacMartin, 2015; Shepherd et al., 2009). Moreover, as with other SRMs, the complexity and uncertainty associated with altering Earth's systems make outcome prediction challenging, creating the potential for unforeseen negative consequences. Finally, the lack of international regulations and governance frameworks for geoengineering introduces regulatory challenges, and the implementation of this technique could potentially lead to unilateral actions with far-reaching global implications and geopolitical tensions (Reynolds, 2019; Shepherd et al., 2009).

Marine Cloud Brightening

(i) What is it?

Solar radiation can be manipulated in several ways through reflective techniques. Among the possible methods, there is marine cloud brightening (MCB), a form of albedo modification. This method leverages stratocumulus clouds layered reflective white clouds, placed on top of the dark ocean to reflect sunlight (Latham et al., 2008). The creation of white clouds is based on the idea of shooting large amounts of sea salt aerosols into low marine clouds from vessels. The particles would attract existing water droplets within the clouds and condense around them, creating fluffier clouds with a larger surface area (Alterskjaer et al., 2013). Whiter clouds with more surface area are also more reflective and can diminish the solar radiation reaching the Earth's surface (Geoengineering Monitor, 2021c). The reflective properties of white surfaces, such as polar ice, are well-documented for their effectiveness in deflecting solar rays (Smedley, 2019).

(ii) Policy integration and implementation examples.

The USA is currently working on an ongoing Marine Cloud Brightening Project (MCBP) founded as 'Silver Lining Inc'. The MCBP is based on an open and international collaboration of scientists and experts who want to advance the understanding of cloud responses to aerosol particles and explore the potential of this geoengineering technique (Marine Cloud Brightening Project, n.d.). Specifically, the MCBP is a collaboration between the University of Washington, the Pacific Northwest National Laboratory (PNNL), a team of engineers from Silicon Valley, the Palo Alto Research Centre and further research partners (Pacific Northwest National Laboratory, n.d.).

The MCBP aims to perform marine cloud brightening by spraying a fine mist of seawater into clouds and announced several outdoor tests based on lab-scale and indoor experiments. However, the first trials were cancelled due to a lack of funding and public opposition to the unknown environmental consequences. In 2019, the MCBP announced the following outdoor experimental research plan:

Stage 1: (Costs: US$ 4–5 M) — > (1) Outdoor dispersion tests at Chico, California; (2) Coastal dispersion test at Moss Landing, California. Goal: to build & test a spray system.

Stage 2: (Costs: US$ 10 M+) — > (3) Coastal cloud spraying at a Californian coastal site; (4) Cloud spraying ~100 km offshore California. Goal: to study basic processes.

Stage 3: (Costs: US$ 10 M+) — > (5) Large-scale cloud brightening experiments: 10,000 km2 area in the NE Pacific, 500–1000 km offshore. The tests will last 1–3 months, each (Geoengineering Monitor, 2021c).

(iii) Strengths/Opportunities.

Maritime Cloud Brightening could be used to mitigate regional warming patterns and specific impacts of climate change. According to MCB, it could be effective at restoring the global climate, despite this technique alone cannot restore the situation to preindustrial conditions (Latham et al., 2012). Compared to other geoengineering techniques, MCB is relatively cost-effective, since materials required for the process, such as seawater or salt particles, are abundant and inexpensive. The effects would also be easier to detect and localize, compared to other geoengineering proposals. In fact, the modification of clouds occurs in specific regions, and the impacts are limited to those areas (Feingold et al., 2024). Another advantage is represented by the fast reversibility of MCB potential effects.

(iv) Weaknesses/Threats.

Current MCB proposals carry several risks and, therefore, need to be further investigated before being applied. According to David Keith's Research Group of Harvard University, even if marine cloud brightening could work, at the same time, it could affect large-scale climate and weather patterns (David Keith's Research Group, 2024). Moreover, marine cloud brightening could only be implemented in limited areas, where the right kinds of clouds exist, which is perhaps only 10 percent of the planet's surface (David Keith's Research Group, 2024). Therefore, several key risks need to be better understood. One of the possible risks is that MCB will cool preferentially in the eastern North/South Pacific and eastern South Atlantic. The potential risks of MCB were highlighted in a 2013 study aimed at evaluating the effectiveness of the technique (ScienceDaily, 2013). The study proved that MCB can counterattack global warming, but it also showed that it could change some environmental conditions making tropical and mid-latitude continents cloudier, wetter, and less prone to drought.

3 Stratospheric, Space-Based and Ground-Based SRM SWOT

3.1 Strengths

Much like ground-based CDR, stratospheric- and space-based SRM holds a set of distinctive strengths. On the one hand, it presents a potential for climate mitigation, deploying solar reflectors or orbital mirrors, countering some global warming effects by curbing solar radiation reaching the Earth's surface. Comparable to ground-based CDR, stratospheric and space-based SRM also ensures global coverage through devices like orbiting mirrors, exerting influence over the entire planet without geographical limitations. Moreover, upon establishing the required infrastructure, swift deployment of stratospheric and space-based SRM systems becomes feasible. For instance, stratospheric aerosol injection (SAI) can be swiftly deployed to counteract climate change effects.

Ground-based SRM methods are largely feasible, such as reflective surfaces or aerosol spraying, which can be implemented using existing technology and infrastructure.

Compared to stratospheric and space-based SRM, ground-based approaches tend to have lower implementation costs and may be more cost-effective and financially accessible.

Moreover, ground-based SRM allows for targeted localized interventions in specific regions, such as urban areas or high-risk regions prone to extreme heat.

3.2 Weaknesses

However, similar to the challenges faced by ground-based CDR, stratospheric and space-based SRM also grapples with certain drawbacks. Developing, launching, and sustaining stratospheric space-based SRM infrastructure, as well as encompassing solar reflectors or orbital mirrors, incurs substantial costs necessitating substantial financial investments. Furthermore, similar to ground-based CDR methods, implementing stratospheric space-based SRM demands advanced technologies for effective sunlight reflection, bringing forth noteworthy technological hurdles (Planetary Sunshade Foundation, 2023). Likewise, uncertainties encircle the long-term environmental, climatic, and ecological impacts of stratospheric and space-based SRM, including SAI, underscoring the need for comprehensive research.

Ground-based SRM methods also face challenges due to their limited range, as they can only impact specific areas or regions rather than achieving global coverage. Moreover, some ground-based SRM techniques, such as aerosol spraying, may have unintended environmental consequences, such as air pollution or changes in local weather patterns. Ground-based SRM methods require ongoing maintenance and monitoring to ensure effectiveness, which can be resource-intensive and may have long-term sustainability challenges.

3.3 Opportunities

If the technology proves viable and safe, space-based SRM, such as deploying solar reflectors or orbital mirrors, could provide a powerful tool for addressing the ongoing climate crisis.

In fact, the development of space-based SRM could offer short-term solutions to address the climate crisis.

Ground-based SRM methods offer scalability and flexibility, allowing for adjustments and adaptations based on local conditions and evolving climate challenges.

Moreover, ground-based SRM approaches, such as incorporating reflective surfaces into buildings or urban planning, can be integrated with existing infrastructure and urban development strategies, highlighting their potential to serve as complementary solutions to broader climate mitigation efforts.

3.4 Threats

Nonetheless, akin to ground-based CDR's challenges, stratospheric and space-based SRM deployment gives rise to several threats. Ethical inquiries emerge, coupled with concerns over manipulating Earth's climate system and the potential for unintended consequences, particularly when utilizing technologies like SAI. Additionally, the absence of a robust international governance framework for stratospheric and space-based SRM, similar to ground-based CDR, may precipitate conflicts, misuse, or unchecked deployment (Lo, 2024). Lastly, stratospheric and space-based SRM, much like ground-based CDR approaches, may encounter public resistance due to environmental apprehensions, possible risks, or perceived diversion from addressing climate change's root causes (Svoboda, 2017). Similarly, ground-based SRM approaches, such as aerosol spraying or altering landscapes, may also face public resistance due to concerns about potential environmental impacts or perceived interference with natural systems. Additionally, the deployment of ground-based SRM methods may encounter regulatory and legal challenges, requiring careful consideration of environmental and public health regulations. The long-term effects and unintended consequences of ground-based SRM methods on ecosystems, weather patterns, and human health are not fully understood and need thorough evaluation, as they may result in unintended consequences (NOAA, 2024).

4 Conclusion

In conclusion, the SWOT analysis conducted for SRM technologies offers an extensive overview of the strengths, weaknesses, opportunities, and threats of the stratospheric, space-based, and ground-based SRM technologies developed by the scientific community as a way to address climate change. In terms of strengths, the SWOT analysis outlines how SRM could leave an immediate impact and produce a rapid cooling effect, while offering a significant amount of flexibility and control. In addition, it could be used as an emergency climate response offering a temporary climate solution. However, there are also downsides connected with SRM technologies. Their limited scope, as they fail to address the root cause of climate change, raises concerns, while their potential to cause unintended consequences introduces complex ethical challenges. This could lead to the possibility of geopolitical tension or even international conflict (McGee et al., 2021). Therefore, on the one hand, SRM has technological limitations and premature and insufficiently researched deployments that could indeed lead to unintended consequences. On the other hand, continued research and investments allow for obtaining a comprehensive scientific understanding and successful development of SRM technologies. Furthermore, it also gives time for achieving a consensus and establishing an appropriate global regulatory framework for climate change mitigation strategies. Finally, it is important to note that potential applications of stratospheric and space-based SRM are still theoretical at this point. SRM necessitates extensive research of potential risks and benefits, and the development of appropriate policy and regulatory frameworks before practical implementation can be considered (Fig. 3.1).

Fig. 3.1 A SWOT for SRM

References

Acker, D., Brauer, U., Fasoulas, S., & Maheswaran, T. (2022). The international planetary sunshade: An umbrella project to Foster International collaboration to mitigate global warming. In *Paper presented at the 73rd international astronautical congress (IAC).*

Akbari, H., Menon, S., & Rosenfeld, A. (2009). Global cooling: Increasing world-wide urban albedos to offset CO2. *Climatic Change, 94*(3–4), 275 286. https://doi.org/10.1007/s10584-008-9515-9

Alterskjaer, K., Kristjánsson, J. E., & Seland, Ø. (2013). Sensitivity to deliberate sea salt seeding of marine clouds—Observations and model simulations. *Atmospheric Chemistry and Physics, 13*(5), 2793–2806.

Arctic Ice Project. (2023). *Arctic Ice Project*. Retrieved December 15, 2024, from https://www.arcticiceproject.org/

Baum, C. M., Low, S., & Sovacool, B. K. (2022). Between the sun and us: Expert perceptions on the innovation, policy, and deep uncertainties of space-based solar geoengineering. *Renewable and Sustainable Energy Reviews, 158*, 112179.

Boyd, P. (2024). *Geoengineering—carbon removal, climate solutions, mitigation*. Encyclopedia Britannica. Retrieved December 15, 2024, from https://www.britannica.com/science/

Chang, D. B., & Shih I. Fu. (1991). *Stratospheric Welsbach seeding for reduction of global warming. United States Patent No. 5003186*. Retrieved December 15, 2024, from https://patents.google.com/patent/US5003186A/en

Climate Overshoot Commission. (2023). *Reducing the risks of climate overshoot*. Edited by Andrew Johnston. Retrieved December 15, 2024, from https://www.climateovershoot.org/reports

Coverstone, V., & Johnson, L. (2015). *Mitigating climate change with earth orbital sunshades*. NASA—Final Report. Preparation of Papers for AIAA Technical Conferences.

David Keith's Research Group. (2024).*Marine cloud brightening*. Harvard University. Retrieved December 15, 2024, from https://keith.seas.harvard.edu/research-marine-cloud-brightening

DeFusco, D. (2011). *Deforestation causes cooling in northern U.S., Canada.* Yale news. Retrieved December 15, 2024, from https://news.yale.edu/2011/03/10/deforestation-causes-cooling-northern-us-canada

European Commission. (2013). *Implications and risks of engineering solar radiation to limit climate change.* CORDIS. Retrieved December 15, 2024, from https://cordis.europa.eu/project/id/226567

European Commission. (2023). *A new outlook on the climate and security nexus: Addressing the impact of climate change and environmental degradation on peace, security and defence.* Joint Communication to the European Parliament and the Council. Retrieved December 15, 2024, from https://ec.europa.eu/commission/presscorner/detail/en/ip_23_3308

European Commission's Group of Chief Scientific Advisors. (2023). *Scoping paper: Solar radiation modification.* Scientific Advice Mechanism. Retrieved December 15, 2024, from https://research-and-innovation.ec.europa.eu/knowledge-publications-tools-and-data/publications/all-publications/scoping-paper-solar-radiation-modification_en

Fachhochschule Graubunden. (2019). *Gletscherschutz wird konkreter.* Retrieved December 15, 2024, from https://www.fhgr.ch/fhgr/medien-und-oeffentlichkeit/medien/medienmitteilungen/gletscherschutz-wird-konkreter/

Feingold, G., Ghate, V. P., Russell, L. M., Blossey, P., Cantrell, W., Christensen, M. W., Diamond, M. S., Gettelman, A., Glassmeier, F., Gryspeerdt, E., Haywood, J., & Zheng, X. (2024). Physical science research needed to evaluate the viability and risks of marine cloud brightening. *Science Advances, 10*(12), eadi8594.

Fuglesang, C., & de Herreros Miciano, M. G. (2021). Realistic sunshade system at L1 for global temperature control. *Acta Astronautica, 186*, 269–279.

Geoengineering Monitor. (2021). *Surface Albedo Modification.* Geoengineering Technology Briefing. Retrieved December 15, 2024, from https://www.geoengineeringmonitor.org/technologies/surface-albedo-modification/

Geoengineering Monitor. (2021a). *Geoengineering Map Explainer: The world map documents geoengineering development Research, experiments and projects.* Retrieved December 15, 2024, from https://www.geoengineeringmonitor.org/2021/05/geoengineering-map-explainer/

Geoengineering Monitor. (2021b). *Stratospheric Controlled Perturbation Experiment (SCoPEx).* Retrieved December 15, 2024, from https://www.geoengineeringmonitor.org/2021/04/stratospheric-controlled-perturbation-experiment-scopex/

Geoengineering Monitor. (2021c). *Marine Cloud Brightening or Cloud Reflectivity Enhancement.* Retrieved December 15, 2024, from https://www.geoengineeringmonitor.org/technologies/marine-cloud-brightening/

International Non-Use Agreement on Solar Geoengineering. (2022). *International Non-Use Agreement on Solar Geoengineering.* Retrieved December 15, 2024, from https://www.solargeoengineering.org/non-use-agreement/

Irvine, P. J., Kravitz, B., Lawrence, M. G., & Muri, H. (2016). An overview of the earth system science of solar geoengineering. *Wiley Interdisciplinary Reviews: Climate Change, 7*(6), 815–833.

Irvine, P. J., Ridgwell, A., & Lunt, D. (2011). Climatic effects of surface albedo geoengineering. *Journal of Geophysical Research, 116*(D24), D24112.

Izrael, Y. A., Zakharov, V. M., Petrov, N. N., Ryaboshapko, A. G., Ivanov, V. N., Savchenko, A. V., Andreev, Y. V., Puzov, Y. A., Danelyan, B. G., & Kulyapin, V. P. (2009). Field experiment on studying solar radiation passing through aerosol layers. *Russian Meteorology and Hydrology, 34*(5), 265–273.

Jones, A. C., Haywood, J. M., Dunstone, N., Emanuel, K., Hawcroft, M. K., Hodges, K. I., & Jones, A. (2017). Impacts of hemispheric solar geoengineering on tropical cyclone frequency. *Nature Communications, 8*, 1382.

Kala, J., & Hirsch, A. L. (2022). Assessing the potential for crop albedo enhancement in reducing heatwave frequency, duration, and intensity under future climate change. *Weather and Climate Extremes, 35*, 100415.

Keith, D. (2020). *Has it come to this? Statement Regarding Financials and Conflict of Interest*. Chapter 13, p. 196.

Keith, D. W., & MacMartin, D. G. (2015). A temporary, moderate, and responsive scenario for solar geoengineering. *Nature Climate Change, 5*(3), 201–206.

Kintisch, E. (2018). *Climate engineering and the law—regulation and liability for solar radiation management and carbon dioxide removal*. Cambridge University Press, edited by Gerrard M. B. & Hester T., Chapter 2, pp. 28–34.

Latham, J., Bower, K., Choularton, T., Coe, H., Connolly, P., Cooper, G., Craft, T., Foster, J., Gadian, A., Galbraith, L., Iacovides, H., & Wood, R. (2012). Marine cloud brightening. *Philosophical Transactions of the Royal Society A: Mathematical, Physical and Engineering Sciences, 370*(1974), 4217–4262.

Latham, J., Rasch, P., Chen, C. C., Kettles, L., Gadian, A., Gettelman, A., Morrison, H., Bower, K., Choularton, T., & Salter, S. (2008). Global temperature stabilization via controlled albedo enhancement of low-level maritime clouds. *Philosophical Transactions of the Royal Society A: Mathematical, Physical and Engineering Sciences, 366*(1882), 3969–3987.

Lawrence, M. G., Schäfer, S., Muri, H., Scott, V., Oschlies, A., Vaughan, N. E., Boucher, O., Schmidt, H., Haywood, J., Scheffran, J., & Schmidt, H. (2018). Evaluating climate geoengineering proposals in the context of the Paris agreement temperature goals. *Nature Communications, 9*(1), 3734.

Lo, J. (2024). *Nations fail to agree ban or research on solar geoengineering*. Climate Home News. Retrieved December 15, 2024, from https://www.climatechangenews.com/2024/02/29/nations-fail-to-agree-ban-or-research-on-solar-geoengineering/

Marine Cloud Brightening Project (MCBP). (n.d.). *About us*. Retrieved December 15, 2024, from https://www.mcbproject.org/about

Massachusetts Institute of Technology. (2022). *MIT researchers propose space bubbles to reverse climate change*. MIT News. Retrieved December 15, 2024, from https://news.mit.edu/2022/space-bubbles-climate-change-0622

McGee, J., Brent, K., McDonald, J., & Heyward, C. (2021). International governance of solar radiation management: Does the ENMOD convention deserve a closer look? *U. of Adelaide Law Research Paper, 2021-19*, 294–305.

Members of the Physical and Biological Community. (2023). *An open letter regarding research on reflecting sunlight to reduce the risks of climate change*. Retrieved December 15, 2024, from https://climate-intervention-research-letter.org/

Moore, J. A., Grinsted, A., Guo, X., Yu, X., Jevrejeva, S., Rinke, A., Cui, X., Kravitz, B., Lenton, A., Watanabe, S., & Ji, D. (2015). Atlantic hurricane surge response to geoengineering. *PNAS, Massachusetts Institute of Technology, 112*(45), 13794–13799.

National Oceanic and Atmospheric Administration. (2024). *Solar radiation modification: NOAA state of the science factsheet*. NOAA Climate.gov. Retrieved December 15, 2024, from https://www.climate.gov/news-features/understanding-climate/solar-radiation-modification

National Research Council. (2015). *Climate intervention: Reflecting sunlight to cool earth*. National Academies Press.

Newcomb, T. (2022, June 29).*MIT scientists propose space bubbles to deflect solar radiation, ease climate change. Popular Mechanics*. Retrieved December 15, 2024, from https://www.popularmechanics.com/science/environment/a40406575/mit-space-bubbles-climate-change/

Oldfield, J. D., & Poberezhskaya, M. (2023). Soviet and Russian perspectives on geoengineering and climate management. *WIREs Climate Change, 14*(4), e829.

Pacific Northwest National Laboratory (PNNL). (n.d.). *Marine Cloud Brightening Project*. Retrieved December 15, 2024, from https://www.pnnl.gov/projects/marine-cloud-brightening

Paddison, L. (2023, February 12). This controversial climate solution could be exactly what the planet needs. *Or it could be a colossal disaster*. CNN. Retrieved December 15, 2024, from https://www.cnn.com/2023/02/12/world/solar-geoengineering-climate-intl/index.html

Patel, T. D., Odoulami, R. C., Pinto, I., Egbebiyi, T. S., Lennard, C., Abiodun, B. J., & New, M. (2023). Potential impact of stratospheric aerosol geoengineering on projected temperature and precipitation extremes in South Africa. *Environmental Research: Climate, 2*(3), 035004.

Planetary Sunshade Foundation. (2023). *State of Space Based Solar Radiation Management*. Retrieved December 15, 2024, from https://www.planetarysunshade.org/

Project Earth. (2008). *Wrapping Greenland0*. Greenland Cover. Retrieved December 15, 2024, from https://projectearth.discovery.com/

Reynolds, J. L. (2019). *The governance of solar geoengineering: Managing climate change in the Anthropocene*. Cambridge University Press.

Rodriguez, S., & Lo, J. (2023). *Mexico plans to ban solar geoengineering after rogue experiment*. Climate Change News. Retrieved December 15, 2024, from https://www.climatechangenews.com/2023/01/18/mexico-plans-to-ban-solar-geoengineering-after-rogue-experiment/

Santamouris, M. (2014). Cooling the cities—A review of reflective and green roof mitigation technologies to fight heat Island and improve comfort in urban environments. *Solar Energy, 103*, 682–703.

Sapinski, J. P., Buck, H. J., & Malm, A. (2021). *Has it come to this?: The promises and perils of geoengineering on the brink*. Rutgers University Press. Retrieved December 15, 2024, from https://doi.org/10.36019/9781978809390

ScienceDaily. (2013, July 10). *Cloud brightening to cool seas can protect coral reefs: Targeted cooling could offer a 50-year breathing space for coral protection*. ScienceDaily. Retrieved December 15, 2024, from https://www.sciencedaily.com/releases/2013/07/130710132443.htm

Shepherd, J. (2009). *Geoengineering the climate: Science, governance, and uncertainty*. The Royal Society.

Siegel, E. (2018, May 26). Ask Ethan: Can we build a sunscreen to combat global climate change? *Forbes*. Retrieved December 15, 2024, from https://www.forbes.com/sites/startswithabang/2018/05/26/ask-ethan-can-we-build-a-sunscreen-to-combat-global-climate-change/

Smedley, T. (2019. February 26). *How artificially brightened clouds could stop climate change*. BBC Future. Retrieved December 15, 2024, from https://www.bbc.com/future/article/20190225-how-artificially-brightened-clouds-could-stop-climate-change

Smith, W. (2020). The cost of stratospheric aerosol injection through 2100. *Environmental Research Letters, 15*(11), 114004.

Smith, W., Visioni, D., Bednarz, E. M., Haywood, J. M., Vuuren, D. P., Cassou, C., & Wordsworth, R. (2018). Impacts of stratospheric aerosol injection geoengineering on global food security. *Nature Food, 4*(1), 83–98.

Smith, W., & Wagner, G. (2018). Stratospheric aerosol injection tactics and costs in the first 15 years of deployment. *Environmental Research Letters, 13*(12), 124001. https://doi.org/10.1088/1748-9326/aae98d

Stiles, L. (2006). *Space sunshade might be feasible in global warming emergency*. Proceedings of the National Academy of Sciences—University of Arizona. Retrieved December 15, 2024, from https://www.arizona.edu/news/space-sunshade-might-be-feasible-global-warming-emergency

Surprise, K. (2019). Stratospheric imperialism: Liberalism, (eco)modernization, and ideologies of solar geoengineering research. *SageJournals, 3*(1), 25–43.

Svoboda, T. (2017). *The ethics of climate engineering: Solar radiation management and non-deal justice*. Routledge.

Tang, A., & Kemp, L. (2021). A fate worse than warming? Stratospheric aerosol injection and global catastrophic risk. *Frontiers Climate, 3*, 720312.

The White House. (2023). *Congressionally mandated research plan and an initial research governance framework related to solar radiation modification*. Retrieved December 15, 2024, from https://www.whitehouse.gov/wp-content/uploads/2023/06/SRM-Report-Final.pdf

Tollefson, J. (2021, March 31).*First solar geoengineering experiment to test releasing particles into the atmosphere*. Nature News. Retrieved December 15, 2024, from https://www.nature.com/articles/d41586-021-00829-1

U.S. National Science Foundation. (2009). *Collaborative Research: Evaluation of Suggestions to Geoengineer the Climate System Using Stratospheric Aerosols and Sun Shading*. NSF Award Search: Award # 0730761. Retrieved December 15, 2024, from https://www.nsf.gov/awardsearch/showAward?AWD_ID=0730761

World Meteorological Organization. (2022).*Executive summary*. Scientific Assessment of Ozone Depletion. GAW Report No. 278. Retrieved December 15, 2024, from https://ozone.unep.org/science/assessment/sap

International Political and Legal Aspects

Governing Carbon Dioxide Removal and Solar Radiation Modification

Joshua Horton

Abstract This chapter explores two distinct approaches, carbon dioxide removal (CDR) and solar radiation modification (SRM), as potential means to counteract climate change, providing additional angles to the SWOT analyses conducted in the previous chapters. CDR methods, both technological and natural, vary in terms of cost, permanence, and removal potential. Governance challenges, such as avoiding moral hazard and ensuring permanence, are critical to implementing large-scale CDR. International governance frameworks like the UNFCCC will play significant roles in shaping CDR policies. Conversely, SRM presents a faster and cheaper but imperfect option, with stratospheric aerosol injection being the most plausible form. Environmental and political risks associated with SRM recommend multilateral decision-making and coordination. Despite research in certain countries, global governance mechanisms remain largely silent on SRM. International political dynamics may influence the trajectory of SRM development and deployment. Ultimately, the fate of both CDR and SRM hinges on international cooperation.

Keywords Carbon dioxide removal (CDR) · Solar radiation modification (SRM) · Governance · International cooperation

1 Introduction

At first glance, carbon dioxide removal (CDR) and solar radiation modification (SRM) may seem similar. Both may appear to involve large-scale interventions in the climate system to reduce global warming. But CDR is not inherently large-scale, and only some forms of SRM qualify as such. And each would address global warming in a fundamentally different way. Indeed, apart from their novelty and their status as climate technologies, CDR and SRM have surprisingly little in common.

J. Horton (✉)
Cambridge, USA
e-mail: joshuahorton533@gmail.com

University of Chicago, Chicago, USA

K.-U. Schrogl et al. (eds.), *The Politics of Geoengineering*, Springer Climate,
https://doi.org/10.1007/978-3-032-10095-5_4

CDR acts on the cause of climate change, i.e., excessive amounts of CO_2 in the atmosphere. In seeking to alter atmospheric concentrations of CO_2, CDR is much more like emissions reductions than SRM—the Intergovernmental Panel on Climate Change (IPCC) defines "mitigation" as emissions cuts plus removals (IPCC, 2022). Like emissions reductions, CDR would be slow to act, requiring decades to influence the climate. To exert such influence, CDR would need to be implemented at large scale. Implementing novel as well as some more familiar CDR methods at large scale would be unprecedented and involve risks. Most of these risks, however, would be local in nature (although cumulative risks could pose broader dangers). CDR is also expensive—more expensive than many readily available emission reduction measures—and large-scale implementation would be very costly, perhaps in the range of $500 billion to $1.6 trillion *per year*.[1]

In contrast, SRM would act on the consequences of climate change, i.e., higher temperatures and associated impacts. Thus, SRM would be much more like adaptation than either CDR or emissions reductions: just as seawalls are built to protect against sea level rise, SRM could be deployed to create what would in effect be a regional or global parasol to reflect sunlight. Unlike CDR (or emissions cuts), SRM could act quickly, generating climate effects within a year or two. Also, unlike CDR, SRM would be cheap—direct costs are estimated to be in the tens of billions of dollars annually (Smith & Wagner, 2018). And SRM would involve regional and global risks, both physical and sociopolitical.

These different profiles present different governance challenges. In theory, existing global governance mechanisms could be adapted to accommodate and promote CDR (although paying for large-scale CDR would require mobilizing massive amounts of additional finance). Existing governance mechanisms are totally unsuited, however, to SRM: brand-new structures would be necessary to manage and oversee this technology.

The next two sections provide more details on the characteristics, policy implications, governance arrangements, and emerging politics of CDR and SRM.

2 CDR

To reach net-zero CO_2, experts estimate that the world will need to remove between 5.5 and 16 $GtCO_2$ per year from the atmosphere by mid-century (Smith et al., 2023). Today, the world removes approximately 2 $GtCO_2$ per year, but almost all of that comes from forestry projects that are vulnerable to reversals and release back to the atmosphere. Achieving net-zero, and thereafter reducing atmospheric concentrations of CO_2 to bring them closer to preindustrial levels, will hence require both developing new CDR methods and deploying them on a large scale.

[1] This rough approximation derives from multiplying the amount of CDR estimated as necessary to reach net-zero—5.5 to 16 $GtCO_2$ per year (Smith et al. 2023)—by $100 per tCO_2, the industry benchmark.

CDR is conventionally divided into two main types, natural and technological (see Table 4.1). Natural CDR, which includes some "nature-based solutions", relies on biomass to absorb and store atmospheric CO_2 (Seddon et al., 2020). Afforestation (planting trees on previously unforested land) and reforestation (planting trees on previously deforested land) are well-known examples of natural CDR. Soil carbon sequestration involves using special agricultural practices, such as no-till farming, to build up soil organic carbon. Biochar is a stable form of carbon produced when biomass is burned under low-oxygen conditions. Additional varieties of natural CDR include agroforestry, "blue carbon" (carbon stored in coastal ecosystems like mangrove forests), and others.

Natural CDR is relatively inexpensive and well-developed and involves comparatively few risks. Further, natural CDR may entail significant co-benefits for local

Table 4.1 CDR Methods

Method	Status (TRL)	Cost (USD/ tCO_2)	Removal Potential ($GtCO_2$/yr)	Risks (examples)	Co-Benefits (examples)
Natural					
Afforestation/ reforestation	8–9	0–240	0.5–10	Reversal of carbon removal	Erosion control, water retention, increased biodiversity
Soil carbon sequestration	8–9	45–100	0.6–9.3	Reversal of carbon removal	Improved soil quality
Biochar	6–7	10–345	0.3–6.6	Unsustainable biomass harvest	Improved crop yields
Agroforestry	8–9	?	0.3–9.4	Reversal of carbon removal	Livelihood enhancement, increased biodiversity
Blue carbon	2–3	?	<1	Reversal of carbon removal	Ecosystem-based adaptation, increased biodiversity
Technological					
DACCS	6	100–300	5–40	CO_2 leakage	–
BECCS	5–6	15–400	0.5–11	Land-use changes, CO_2 leakage	Energy production
OAE	1–2	40–260	1–100	Ecosystem impacts, impacts from mining	Reduced ocean acidification

Source IPCC AR6 WGIII Table TS.7
Note Technology readiness level (TRL) refers to the relative technical maturity of a given technology on a scale from 1 (basic principles observed) to 9 (actual system proven)

communities, for example, growing forests may reduce erosion, increase water retention, and provide various livelihood opportunities. CO_2 stored in biomass, however, is vulnerable to reversals. Forest fires, for instance, burn trees and release their stored CO_2 straight back to the atmosphere. Alternatively, changes in agricultural practices can disrupt soils and release their stored CO_2. CO_2 storage provided by natural CDR is thus non-permanent. Moreover, natural CDR is limited in terms of the absolute quantity of CO_2 it can remove from the atmosphere.

Technological CDR, by contrast, relies primarily on technology—often industrial-scale—to remove CO_2 from the atmosphere. The most discussed technological CDR option today is direct air carbon capture and storage (DACCS), which uses chemicals to "scrub" CO_2 directly from the air and then stores it permanently underground. Bioenergy carbon capture and storage (BECCS) involves burning biomass to generate heat and/or power, capturing the resulting CO_2 at the smokestack, and storing it underground; since the burnt biomass contained CO_2 taken from the air via photosynthesis, capturing CO_2 post-combustion indirectly removes it from the atmosphere. Ocean alkalinity enhancement (OAE) would entail dispersing alkaline materials like lime in seawater, allowing the ocean to absorb more atmospheric CO_2 while reducing ocean acidification.

Technological methods are expensive and underdeveloped and may pose local risks. Supplying biomass to fuel BECCS, for example, may drive land-use changes resulting in higher food prices and biodiversity loss. CO_2 stored in geologic formations for BECCS and DACCS, if not properly managed, could leak into groundwater or back to the atmosphere. OAE could negatively affect marine ecosystems. However, assuming technological CDR is well-planned and operated based on established protocols, it would provide permanent storage and have the potential to remove very large amounts of CO_2 from the atmosphere.

Currently, support for natural CDR is more widespread than support for technological CDR. This is attributable to several factors: a general preference for "nature" over "technology" in responding to climate change; greater familiarity with existing natural CDR methods compared to more novel technological methods; and fewer risks and more co-benefits associated with natural CDR.[2] But from a climate perspective, natural CDR will not be sufficient on its own: its limited removal potential makes natural CDR incapable of offsetting continuing emissions and bringing down atmospheric CO_2 concentrations, while its non-permanent storage risks returning previously absorbed CO_2 back to the atmosphere. Thus, future CDR must involve both natural and technological methods.

Governing CDR will require addressing multiple challenges. First, given the scale of CDR required to reach net-zero and beyond, governments will need to promote innovation (particularly for technological options), support demonstration, and spur deployment. Innovation can be promoted through public funding for research and development (R&D). Demonstration projects can be supported through

[2] All CDR methods—even those typically regarded as "natural"—involve the use of technology. Afforestation and reforestation, for example, depend on a host of technologies to grow and transplant tree seedlings.

public-private partnerships and "de-risking". Eventually, large-scale deployment can be implemented directly by governments, purchased through public procurement, incentivized through tax credits, or required by imposing legal obligations on private actors.

Technology-push and demand-pull policies to catalyze breakthroughs, cut costs, and drive implementation have been pursued primarily in the US and Europe. Worldwide public funding for CDR research, development, and demonstration (RD&D) totaled approximately $4.1 billion from 2010 through 2022 (Smith et al., 2023). Most of that—$3.5 billion—is tied to the US Department of Energy Regional Direct Air Capture Hubs program, which aims to support construction of four demonstration hubs across the country. So-called 45Q tax credits in the US offer a range of incentives to CDR project developers, including $180 per tonne of CO_2 removed and stored in geologic formations using DACCS.

Second, to facilitate this, governments will need to provide for reliable measurement, reporting, and verification (MRV). Measurement involves quantifying removals (including on a life-cycle basis) and assessing co-benefits. Reporting involves compiling and communicating this information to authorities and the public. And verification entails validating the accuracy of reported information. Robust project-level accounting promotes transparency and accountability and allows for the generation of tradable removal credits.

Third, governments will need to guard against the risk of reversal posed by non-permanent natural CDR. Several options are available. One is to categorize credits sourced from natural CDR methods as "temporary", with specified expiration dates; this approach was applied to forestry credits under the Kyoto Protocol's Clean Development Mechanism (CDM). Another is to require buffer pools for natural CDR credits: in the event of a reversal, lost credits would be replaced by additional unused credits set aside in the pool. Strong MRV will help ensure that reversals are detected and addressed.

Fourth, governments will also need to promote co-benefits associated with natural CDR. For instance, governments could explicitly recognize and compensate for adaptation benefits provided by forestry projects. Similarly, governments could pay additionally for ecosystem services like coastal protection provided by mangrove restoration projects. Effective MRV can track and document co-benefits.

The EU is currently considering a Carbon Removal Certification Framework (CRCF) that lays out an MRV system incorporating permanence safeguards and requirements for co-benefits (European Commission, 2022). The CRCF would apply to natural and technological CDR. Methodologies for individual methods are to be developed based on "QU.A.L.ITY" criteria: QUantification (measurement), Additionality (removals would not have been implemented without opportunities provided by certification), Long-term storage (permanence), and sustainabilITY (co-benefits). Registered certification bodies will verify that removals meet criteria specified in relevant methodologies and issue certificates, which may eventually be tradable as offsets within the EU Emissions Trading System (ETS). Observers anticipate some version of the CRCF will be adopted in the next few years.

Fifth, while advancing CDR development and deployment, governments must ensure that CDR does not come at the expense of emissions reductions. Many observers have raised concerns that the prospect of CDR may weaken efforts to cut emissions; this is sometimes referred to as "moral hazard". This is already occurring, for instance, in the case of corporate carbon offsetting using "phantom credits" generated by suspect forestry projects (Greenfield, 2023). One way to reduce this risk would be for policymakers to adopt separate targets for emissions reductions and carbon removal (McLaren et al., 2019).

Finally, other risks posed by CDR, especially by technological methods, must be reduced and managed. Biomass intended for BECCS, for example, could be required to meet sustainable sourcing standards to protect against environmental and social externalities. Permits for building or operating BECCS or DACCS facilities could be made conditional on agreeing to accept liability in the event of CO_2 leakage from geologic formations.

At the global level, governance of CDR is currently restricted to exploratory work being carried out by the Paris Agreement "Article 6.4 Mechanism Supervisory Body". The Paris Agreement's Article 6.4 Mechanism is intended to replace the CDM, and the Supervisory Body is tasked with setting it up. The Supervisory Body is now considering whether and, if so, how, removal credits should be incorporated in a future Article 6.4 Mechanism. Because the EU ETS was the main market for CDM offset credits and may play a similar role for credits issued under the Article 6.4 Mechanism, any rules on CDR put forward by the Supervisory Body are likely to reflect provisions contained in the CRCF.

Given its high costs and the need for large-scale deployment, the global politics of CDR will likely be dominated by debates over who should pay for it. In the first instance, discussions will center on which countries should pay. Arguments are likely to emphasize historical responsibility and ability to pay as bases for determining the distribution of costs across countries (Pozo et al., 2020).

Within countries, decision-makers will have to wrestle with similar issues. Government implementation or procurement would place the burden on taxpayers. Alternatively, specific actors or sectors could be singled out to pay the costs of CDR. For example, governments could assign "carbon takeback obligations" to fossil fuel companies requiring them to remove and store CO_2 from the atmosphere (Bednar et al., 2021; Jenkins et al., 2021). Large-scale deployment may necessitate the use of market-based mechanisms.

3 SRM

Two main types of SRM have been proposed. Stratospheric aerosol injection (SAI) would involve releasing limited amounts of aerosols in the upper atmosphere (stratosphere) to reflect a small fraction of incoming sunlight back to space, thereby cooling the planet. Scientists are confident that implementing SAI would reduce temperatures because it would mimic the known cooling effects of sulfate aerosols released by large volcanic eruptions; the 1991 eruption of Mount Pinatubo in the Philippines, for example, reduced global temperatures by approximately 0.5 °C in the years that immediately followed. Familiarity with such natural analogs makes sulfates the type of aerosol most considered for use in SAI and makes SAI the most plausible form of SRM.

Marine cloud brightening (MCB) would entail spraying seawater into low-lying marine clouds to whiten them and reflect more sunlight. MCB would imitate the cooling effects associated with ship tracks created by exhaust from oceangoing vessels. Compared to SAI, MCB is much less studied.

What sets these technologies—especially SAI—apart from other approaches is that they would be relatively fast and relatively cheap. They would be fast because the cooling effects resulting from their deployment would be observable within a year or two. And they would be cheap because the direct cost of implementing them on a global scale would be in the tens of billions of dollars per year.

But they would also be imperfect—they could not perfectly compensate for the warming caused by GHGs—because greenhouse warming and sunlight reflection are different physical processes. Using SRM could move a warmer world closer to preindustrial conditions, but it could not recreate those conditions—SRM and emissions cuts are not substitutes. Some aspects of a geoengineered climate, for example, regional precipitation patterns, would depart from the historical average in ways that might result in harms, even if overall benefits outweighed risks by a significant margin.

Indeed, the available evidence from climate models consistently shows that, under certain conditions, SRM could produce considerable net benefits for people and nature. Specifically, deploying SRM could curb and curtail warming (including heat-waves), moderate floods and droughts, weaken tropical storms, and limit sea level rise, all risks that are expected to worsen with climate change (NASEM, 2021). Further, since these and similar impacts will harm developing countries disproportionately, these countries stand to benefit disproportionately if SRM is used appropriately (Harding et al., 2020).

Yet SRM would also entail significant risks. Using sulfate aerosols in SAI would delay recovery of the ozone layer. Novel climate features attributable to SRM such as altered precipitation patterns could cause damage. If a large-scale deployment were halted abruptly while atmospheric GHG concentrations were elevated, a "termination shock" could ensue in which temperatures would rebound at a faster and more destructive rate than would have occurred had SRM never been implemented. As with CDR, the perceived availability of SRM might undermine the drive to reduce emissions, posing a moral hazard. And the inexpensive nature of the technology might permit individual countries (or small groups of countries) to deploy SRM unilaterally, imposing their preferred global temperature on the entire world.

In practice, the global balance—and distribution—of benefits and risks from SRM would depend largely on how the technology was used. Using SRM is not a binary choice—it is not a question of either using it or not using it. Instead, SRM could be used in a variety of ways to pursue a range of goals, from "shaving the peak" off warming to make time for decarbonization, to fully offsetting warming to return the world to preindustrial temperatures. And design goes beyond the question of "where to set the global thermostat": deployments could vary in terms of temperature and/or precipitation goals, rates of ramp-up, duration, conditions for cessation, and rates of phasedown. Further, any given deployment could be fine-tuned by adjusting aerosol release in terms of hemisphere, latitude, season, etc. to reduce the extent of harmful novel climate features. As a rule, however, the risks associated with SRM increase roughly in proportion to the size of the intervention being considered, which means that smaller interventions tend toward a more favorable balance of benefits to risks. Early research indicates that moderate, carefully controlled deployments—designed, for example, to slow the rate of warming by half—could deliver substantial global benefits with comparatively low risks (e.g., Irvine et al., 2019).

The potential benefits and risks of SRM would both be subject to a significant degree of uncertainty. The inherently chaotic nature of the climate system makes precise predictions about its future impossible to make. Such uncertainty, however, applies to future climates both with and without SRM (which are forecast using the same computer models) and to negative as well as positive impacts of SRM. In addition, this uncertainty is bounded by knowledge about past climates and past events (like large volcanic eruptions).

Because of the high stakes involved in decisions about whether and how to deploy SRM, arrangements for governing the technology are critically important. The governance of SRM would need to perform several functions. First and foremost, countries will need to decide about deployment. As discussed above, this will not be a binary choice, but rather a choice between different probable but uncertain climate futures, some with SRM and some without. Choosing between these, and thus deciding whether to use SRM, will entail assessing "risk-risk tradeoffs" between worlds experiencing climate change and worlds experiencing climate change but also implementing SRM.[3] This choice will be fundamentally and unavoidably political.

The political, and global, nature of this decision makes it fit for a multilateral forum, within which governments could vote on how to proceed. By the same logic, however, such a decision is unfit to be made on a unilateral basis, and governance mechanisms may need to be built to constrain efforts to implement SRM in the absence of international consensus. Alternatively, such constraints may have to be more geopolitical than institutional.

Second, if the international community decided to implement SRM, there would be a need to ensure that using SRM does not weaken emissions reductions. As noted earlier, SRM and emissions cuts are not substitutes, and yet behavioral and/or political drivers may work to encourage this view. Guarding against moral hazard might be accomplished, for example, by making voting rights on SRM deployment decisions contingent on delivering substantial, verifiable emissions reductions (Parson, 2014).

Third, since deployment could last for decades or even centuries, an SRM governance framework would need to provide for a minimum level of continuity and

[3] When thinking about SRM, a common mistake is to compare a future world with more climate change plus SRM to the present world with less climate change and no SRM (or alternatively to the preindustrial world), and erroneously attribute future climate impacts to SRM.

robustness to avoid premature or unplanned cessation and the risk of termination shock. This might be partly achieved through procedural or bureaucratic mechanisms designed to prevent high-level political authorities from making hasty or ill-considered decisions. Redundancies, both institutional and infrastructural, might also help protect deployment against unanticipated shocks.

Fourth, while some decision-making processes may need to be slowed down, others may need to be insulated against political interference or shielded from overly burdensome procedural requirements so that time-sensitive decisions can be made, and appropriate actions taken. For instance, a large volcanic eruption could amplify the effects of any preplanned deployment operations, such that system operators might need to halt deployment temporarily to avoid an undesired outcome. Technical experts must have sufficient room to maneuver when addressing urgent or specialized matters. Such room could be maintained by explicitly delegating a circumscribed set of authorities to specific operators to be exercised under defined conditions.

Finally, a credible compensation mechanism would be necessary. Countries may claim to have suffered harm from SRM, for example, due to drought caused by altered regional precipitation. If such a claim were made, it would first need to be evaluated, and if it were regarded as valid, then countries would need to be compensated. The inherently probabilistic nature of the climate would appear to make liability mechanisms premised on deterministic causal relationships unsuited to resolving claims related to SRM. An insurance mechanism may therefore be a more appropriate tool for addressing and providing compensation for damages from SRM (Horton and Keith, 2019).

While the scale of the governance challenges posed by SRM is thus immense, current global governance of SRM is virtually non-existent. The centerpiece of international climate policy, the UNFCCC, has so far said nothing about the technology. The most significant development to date is a 2010 decision by the parties to the Convention on Biological Diversity (CBD) calling for no SRM in the absence of effective global governance, except for small-scale research.[4] This decision, however, is not legally binding, and the US is not a party to the CBD.

This governance gap is ultimately attributable to the controversy surrounding SRM. Concerns about moral hazard have focused on some specific interrelated issues including, among others, the possibility that fossil fuel interests might exploit SRM to weaken climate action; the excuse SRM might provide decision-makers to avoid deep decarbonization; and the opportunity SRM might provide developed countries to evade their historical responsibility for climate change. Other risks associated with SRM, like termination shock and the threat of unilateral deployment, also provoke unease. Further, the supposedly "unnatural" status of SRM has fueled objections that it would be an inappropriate tool for climate protection.[5]

---

[4] To be more specific, this CBD decision asked parties to consider prohibiting "geo-engineering", including SRM as well as CDR.

[5] It is not always clear or obvious what some commentators mean when they describe SRM as "unnatural"—typically it appears intended to convey some combination of anthropogenic, deliberate, and perturbative.

Fueled by such concerns, SRM was in effect a taboo subject within the research community until the late 2000s, and even today it receives only 0.1 percent of the support provided to other areas of climate science (this figure derives from data from Overland & Sovacool, 2020; Horton, 2022.). Research is most advanced in the US, Germany, China, and Australia, but none of these countries has a national research program, and no outdoor experiments have been conducted thus far (Horton et al., 2023). With little research being carried out, and little incentive to take up controversial issues in the absence of clear political benefits, the policy community has largely ignored the topic. In turn, this lack of scientific and policy engagement has contributed to—and been reinforced by—a widespread lack of public awareness about SRM. Limited scientific, policy, and public interest in the technology has meant limited demand for global governance.

Yet three things, individually or in combination, may change this in the coming years. First, despite a longstanding assumption that developing countries would be strongly opposed to SRM, emerging evidence suggests there may be latent support in the Global South for expanded research and increased consideration (Sovacool et al., 2023). This possibility should not be surprising: since developing countries are likely to suffer the most from higher temperatures, they are also likely to benefit the most from reducing them.

Second, as strategic competition between the US and China heats up, it is increasingly likely that SRM, like other science and technology issues, will get swept up in the rivalry. One possibility is that SRM comes to be viewed as a prestige technology signifying superpower status. Growing interest in SRM among developing countries, whose support both the US and China are seeking, may further raise the profile of SRM in competition between the two countries.

Third, the idea of an international moratorium on large-scale SRM interventions is emerging as a focal point in discussions about possible near-term governance (Climate Overshoot Commission, 2023). A moratorium would impose a temporary ban on most SRM interventions (presumably with an exception for small-scale research) while specifying the conditions under which it might be lifted. The CBD decision noted above falls short of the type of legally binding, broadly subscribed moratorium which is increasingly being debated among influential stakeholders.

4 Conclusion

Looking forward, as noted earlier, the global politics of CDR will likely be dominated by debates over who should pay for it. This is not dissimilar from ongoing debates about who should pay for emissions reductions and adaptation, which suggests that comparable patterns of under-provision may plague carbon removal at the global level. Governments may be tempted to place the financial burden of CDR on fossil fuel companies by imposing carbon takeback obligations. The fossil fuel industry—and petrostates—would almost certainly resist, especially in a context in which low-

and no-carbon energy sources are increasingly competitive with—if not cheaper than—coal, oil, and gas.

A political bargain is, however, conceivable. As discussed above, large amounts of technological CDR such as DACCS and BECCS will be necessary. These methods involve storing CO_2 underground, which requires transport, injection, and storage management capabilities. These skills and expertise are concentrated in the fossil fuel industry, which has substantial experience building and operating pipelines, compressing CO_2 for transport and injecting it underground (for enhanced oil recovery), and characterizing and managing underground geologic formations. Governments and fossil fuel companies might strike a deal in which the latter accept takeback obligations in exchange for a central role in implementing DACCS, BECCS, and similar technological CDR methods, providing the industry with a crucial lifeline in an era of decarbonization. Such bargains could be struck at national and international levels, helping to resolve the burden-sharing issue. (They would also attract criticism for appearing to benefit fossil fuel companies).

In contrast, the global politics of SRM will not be driven by cost concerns. Indeed, "the global politics of SRM" barely exists—debates in the near future will likely center on whether SRM should be on the international policy agenda in the first place, and if so, how it should be framed. Pressures may come from above and below. From above, policy entrepreneurs may seek to "normalize" the topic in international discussions. Conflicts over agenda-setting may crystallize around the idea of a moratorium. Some actors are trying to prevent normalization of SRM as a climate policy option (e.g., Biermann et al., 2022). From below, developments at national level—in particular, the possible establishment of national SRM research programs—may raise its global profile. The US is most likely to establish such a program. If that were to happen, there are reasons to believe that Europe and China would follow suit (Horton et al., 2023). Were SRM to get on the global agenda, the path by which it arrived would invariably shape subsequent political debates about, for example, outdoor experiments, global governance, and, if deployment were considered, where to set the "global thermostat".

Neither CDR nor SRM should be pursued on its own. CDR *should* supplement emissions reductions in slowing the increase of, and ultimately reducing, the atmospheric stock of CO_2. SRM *could* complement emissions cuts, CDR, and adaptation by reducing otherwise unavoidable climate impacts. Thus, both approaches should be understood as potential elements of a comprehensive climate policy portfolio, featuring different attributes but working toward the same overall goal of climate risk reduction.

References

Bednar, J., Obersteiner, M., Baklanov, A., Thomson, M., Wagner, F., Geden, O., Allen, M., & Hall, J. W. (2021). Operationalizing the net-negative carbon economy. *Nature, 596*, 377–383.

Biermann, F., Oomen, J., Gupta, A., Ali, S. H., Conca, K., Hajer, M. A., Kashwan, P., Kotze, L. J., Leach, M., Messner, D., Okereke, C., Persson, A., Potocnik, J., Schlosberg, D., Scobie, M., & VanDeveer, S. D. (2022). Solar geoengineering: The case for an international non-use agreement. *WIREs Climate Change, 13*, e764.

Climate Overshoot Commission. (2023). *Reducing the Risks of Climate Overshoot*. Retrieved December 15, 2024, from https://www.overshootcommission.org/_files/ugd/0c3b70_bab3b3c1cd394745b387a594c9a68e2b.pdf

European Commission. (2022). *Proposal for a Regulation of the European Parliament and of the Council Establishing a Union Certification Framework for Carbon Removals*. Retrieved December 15, 2024, from https://climate.ec.europa.eu/system/files/2022-11/Proposal_for_a_Regulation_establishing_a_Union_certification_framework_for_carbon_removals.pdf

Greenfield, P. (2023). *Revealed: More than 90% of rainforest carbon offsets by biggest certifier are worthless, analysis shows*. Guardian. Retrieved December 15, 2024, from https://www.theguardian.com/environment/2023/jan/18/revealed-forest-carbon-offsets-biggest-provider-worthless-verra-aoe

Harding, A. R., Ricke, K., Heyen, D., MacMartin, D. G., & Moreno-Cruz, J. (2020). Climate econometric models indicate solar geoengineering would reduce inter-country income inequality. *Nature Communications, 11*, 2801. https://doi.org/10.1038/s41467-019-13957-x

Horton, J. B., Brent, K., Dai, Z., Felgenhauer, T., Geden, O., McDonald, J., McGee, J., Schneuit, F., & Xu, J. (2023). Solar geoengineering programs on National Agendas: A comparative analysis of Germany, China, Australia, and the United States. *Climatic Change, 176*, 83.

Horton, J. B., & Keith, D. W. (2019). Multilateral parametric climate risk insurance: A tool to facilitate agreement about deployment of solar geoengineering? *Climate Policy, 19*(7), 820–826.

Intergovernmental Panel on Climate Change (IPCC). (2022). *Climate Change 2022: Mitigation of Climate Change*. Contribution of Working Group III to the Sixth Assessment Report of the Intergovernmental Panel on Climate Change. Cambridge University Press. Retrieved December 15, 2024, from https://www.ipcc.ch/report/ar6/wg3/

Irvine, P., Emanuel, K., He, J., Horowitz, L. W., Vecchi, G., & Keith, D. (2019). Halving warming with idealized solar geoengineering moderates key climate hazards. *Nature Climate Change, 9*(4), 295–299.

Jenkins, S., Mitchell-Larson, E., Ives, M. C., Haszeldine, S., & Allen, M. (2021). Upstream decarbonization through a carbon takeback obligation: An affordable backstock climate policy. *Joule, 5*(11), 2777–2796.

McLaren, D. P., Tyfield, D. P., Willis, R., Szerszynski, B., & Markusson, N. O. (2019). Beyond net-zero: A case for separate targets for emissions reduction and negative emissions. *Frontiers in Climate, 1*, 4.

National Academies of Sciences, Engineering, and Medicine (NASEM). (2021). *Reflecting sunlight: Recommendations for solar geoengineering research and research governance*. National Academies Press.

Parson, E. A. (2014). Climate engineering in global climate governance: Implications for participation and linkage. *Transnational Environmental Law, 3*(1), 89–110.

Pozo, C., Galan-Martin, A., Reiner, D. M., Mac Dowell, N., & Guillen-Gosalbez, G. (2020). Equity in allocating carbon dioxide removal quotas. *Nature Climate Change, 10*(7), 640–646.

Seddon, N., Chausson, A., Berry, P., Girardin, C. A. J., Smith, A., & Turner, B. (2020). Understanding the value and limits of nature-based solutions to climate change and other global challenges. *Philosophical Transactions of the Royal Society B, 375*(1794), 20190120.

Smith, S.M., Geden, O., Minx J.C., & Nemet, G. F. (2023). *The State of Carbon Dioxide Removal*. Retrieved December 15, 2024, from https://static1.squarespace.com/static/633458017a1ae214f3772c76/t/63e3d4602156db24bc18c91c/1675875445298/SoCDR-1st-edition.pdf

Smith, W., & Wagner, G. (2018). Stratospheric aerosol injection tactics and costs in the first 15 years of deployment. *Environmental Research Letters, 13*(12), 124001.

Sovacool, B. K., Baum, C. M., Low, S., Roberts, C., & Steinhauser, J. (2023). Climate interventions, just transitions, and sustainable development: Global public perceptions. *Global Environmental Change, 82*, 102722.

[illegible]

Asia-Pacific Perspectives on Solar Geoengineering

Masahiro Sugiyama

Abstract This chapter reviews research, governance activities, and public perceptions related to solar geoengineering with a focus on the Asia-Pacific region. Despite significant advances in research and governance, such activities are highly concentrated in high-income countries and biased toward the Global North. An emerging literature has revealed differences across geographies in terms of public perceptions, cultural traditions, and attitudes toward large-scale engineering. Some studies showed that the publics in the Global South are more open to the consideration of solar geoengineering than those from the Global North, though they are all cautious about research and deployment and in favor of global governance. Furthermore, Asia-Pacific regions have been pursuing various regional intervention research projects. To reflect diverse perspectives, this chapter argues for more inclusive research and governance discussion.

Keywords Solar radiation modification · Non-Western perspectives · Research collaboration · International governance · Weather modification · Public engagement

1 Introduction

Solar geoengineering, or solar radiation modification (SRM), is intended to cool the global climate system to counteract the ever-worsening impacts of anthropogenic climate change, and its effect would be felt across the globe, if implemented. Because of its global reach, the consideration of SRM should reflect various global perspectives. In other words, it would be desirable to engage with about 200 member states of the United Nations and their citizens and stakeholders. And yet, most of research, governance discussion, stakeholder consultation, and public engagement conducted to date has been confined to North America and Western Europe. This is even more alarming as the pace of deliberation is gathering momentum. The United States (US)

M. Sugiyama (✉)
Institute for Future Initiatives, The University of Tokyo, Tokyo, Japan
e-mail: masahiro@ifi.u-tokyo.ac.jp; masahiro_sugiyama@alum.mit.edu

K.-U. Schrogl et al. (eds.), *The Politics of Geoengineering*, Springer Climate,
https://doi.org/10.1007/978-3-032-10095-5_5

White House recently issued a congressionally mandated report on a possible US research program (White House Office of Science and Technology Policy (OSTP), 2023). The European Union developed a report on climate and security nexus (High Representative of the Union for Foreign Affairs & Security Policy, 2023), indicating its support for international assessments on SRM as well as discussions on SRM governance. Countries led by Switzerland submitted a draft resolution on SRM to the 2024 United Nations Environment Assembly, requesting an assessment of SRM by the United Nations Environment Programme, though the proposal was eventually withdrawn because of wide opposition (Biermann & Gupta, 2024; McLaren & Corry, 2025).

Perspectives from outside North America and Western Europe are important for two reasons. First, climate change disproportionately impacts those from the Global South or vulnerable countries (IPCC, 2022), and it is crucial to listen to these communities if SRM were to be used to ameliorate the impacts of climate change. Second, emerging economies might be tempted to deploy solar geoengineering in the case of an accelerating climate catastrophe since different countries have fundamentally different interests in SRM (Ricke et al., 2013). How is SRM being discussed or researched in those jurisdictions? These will be key inputs for the consideration of possible international collaborations, as argued by National Academies of Sciences, Engineering, and Medicine (2021), Sugiyama et al. (2017a, 2017b), White House Office of Science and Technology Policy (OSTP) (2023).

It is important to note, at the outset, that there is a diversity of opinions, even in the Global South. It is not fair to bundle diverse countries into a single grouping. For instance, an author for the Intergovernmental Panel on Climate Change (IPCC), Prof. Chukwumerije Okereke (a Coordinating Lead Author for the IPCC 6th assessment report), wrote a passionate plea against SRM in The New York Times (Okereke, 2023). He argued that SRM is not in the interest of Africa because of a multitude of risks and rejected both development and deployment of SRM. In response to this, a group of African scientists involved in the Degrees Initiative, a non-profit organization promoting SRM science in developing countries countered this assertion, saying that vulnerable communities from the Global South might suffer or benefit from SRM, and that further research is warranted (Abiodun et al., 2023). This is just one example of disagreements about this option in the Global South. A more diverse set of opinions can be identified. The ETC Group, which has offices in the Philippines and Mexico, has fiercely opposed to SRM, with a Hands off Mother Earth campaign since 2010 (ETC Group, n.d., 2010), for instance.

This chapter aims to briefly review the perspectives outside North America and Western Europe, with a particular focus on large greenhouse gas emitters in the Asia-Pacific region because of their potential scientific and technical prowess.

The chapter starts by laying out the need for global conversation on SRM. It then reviews the ongoing discussions about SRM in three Asia-Pacific countries with major emissions: China, India, and Japan. The chapter concludes by contrasting the perspectives in the three countries against those from North America and Western Europe. Though Japan is sometimes categorized as part of the West, it is still instructive to examine it because Japan sometimes shows a unique position. It concludes by emphasizing the heterogeneity even within a country and noting that global governance needs to reflect various perspectives from across the world.

2 Solar Geoengineering and Global Conversation

Many calls have been made to begin a global conversation around this controversial technology (or more precisely putative technology) (Carr et al., 2013; Rahman et al., 2018), but the progress has been rather slow.

There have been a number of initiatives to "open up" (Fiorino, 1990; Stirling, 2008) the discursive space to include the voices from the Global South. SRMGI (now the Degrees Initiative) has conducted a number of workshops on solar geoengineering across the globe; see (African Academy of Sciences, 2013) for an African perspective. Winickoff et al. (2015) conducted a workshop on solar geoengineering in partnership with SRMGI, with 39 participants out of the total of 45 from the Global South. The workshop led to critical insights such as the need for international accountability of small-scale outdoor research and for involving the Global South in SRM research and governance. Also Beyerl & Maas (2014) discussed a 2014 workshop in Fiji, in which participants from small Pacific island countries discussed climate engineering.

More recently, the Degrees Initiative (renamed after SRMGI) has been funding research projects in which researchers from the Global South work together with those from the Global North to understand potential risks and benefits of SRM. Carnegie Climate Governance Initiative (C2G) has been fostering global and inclusive conversations on SRM and carbon dioxide removal around the world, with policymakers, researchers, and stakeholders (see the comments from the C2G Director and colleagues: Pasztor, 2021; Pasztor et al., 2017). In 2023, the Alliance for Just Deliberation on Solar Geoengineering (https://sgdeliberation.org/) began its operation to facilitate global deliberation (Temple, 2023).

Besides, some studies were conducted to examine the opinions from the Global South with various methods and respondents: a six-country survey with representative samples (Visschers et al., 2017), a survey with undergraduate students in six Asia-Pacific countries (Sugiyama et al., 2020), a survey with stakeholders (Hussain et al., 2023), interviews in the South Pacific, Sub-Saharan Africa, and the North American Arctic (Carr & Yung, 2018), and a survey in four Asia-Pacific countries (Sugiyama et al., 2025). Most recently Baum et al. (2024) and Brutschin et al. (2024) conducted an extensive survey study of nationally representative samples from 30 countries including many jurisdictions which had never been investigated before, which has set a new benchmark (though this study examined SRM and CDR perceptions simultaneously). These studies show that the awareness of SRM is generally low, and that the respondents from the Global South are likely to show more openness to SRM than the counterparts from the Global North, and that there is a broad agreement on governance. A survey with international climate negotiators and scientists was also conducted (Dannenberg & Zitzelsberger, 2019), which revealed that the countries with lower per-capita GDP tend to support SRM as an option to address climate change, though SRM receives a lower level of priority than mitigation and adaptation.

Opening up the conversation is not only about involving developing, low-income countries but also about engaging emerging economies. Various workshops and events have been organized in different places, including Singapore, Senegal, South Africa, Ethiopia, Fiji, and Tanzania. These events were mostly organized by the Solar Radiation Management Governance Initiative (SRMGI) and their collaborators (see Winickoff et al., 2015 for a review until 2015). The present author and collaborators held a workshop in Japan, inviting experts from the Asia-Pacific region (Sugiyama et al., 2017a, 2017b). Delina (2020) reported a 2019 workshop held in Bali, Indonesia, which was also helped by the SRMGI. More recently, the Degrees Initiative (which was renamed from SRMGI) held a workshop in Mexico in April, 2023 (Degrees Initiative, 2023).

While these efforts are laudable, there remains a substantial problem. One important point is that because SRM used to be a taboo for a long time before Crutzen's (2006) publication, only few official venues for multilateral discussion have dealt with this topic. The Convention on Biological Diversity and London Convention and Protocol have had discussion on this, but the main forum on climate change, i.e., the United Nations Framework Convention on Climate Change, and the scientific assessment, Intergovernmental Panel on Climate Change, rather avoided or downplayed this topic (Reynolds, 2019; Reynolds et al., 2022).

There are some problems with the current situation. First, there is little information about SRM in non-English languages and this precludes wide debate. (Most of official United Nations documents are available in the six UN official languages whereas the academic literature is often available only in English.) Second, states and stakeholders are not exposed to the detailed discussion, and they have not formed detailed position on SRM. Third, there is no formal, international collaboration framework even for observations and modeling, except for the Geoengineering Model Intercomparison Project (GeoMIP), which has been supported by voluntary work of individual researchers.

Nevertheless, there are some activities (perhaps indirectly) related to SRM in various countries. Many countries have had a history of weather modification and artificial rainmaking, whose history can be related to that of SRM (Fleming, 2010). More than 50 countries have operational programs of weather modification (fog dispersal, rain and snow enhancement, etc.) (Munoz, 2017), including the United States, China (Bluemling et al., 2020; Chien et al., 2017) and Thailand (Chirapant, 2022) (see also https://map.geoengineeringmonitor.org/ for a list of weather modification projects across the globe). The rest of the chapter tries to summarize the debate around SRM.

3 Country Perspective: China

The following, brief discussion on China's policy draws on (Horton et al., 2023), which reviewed SRM policy development using the two frameworks (role theory from international relations and multiple streams approach from policy sciences).

China has been engaged with research on SRM. The most notable program is "Mechanism and Impact of Geoengineering" (sometimes called China Geoengineering Research Program), which was conducted from 2015 and 2019 jointly at Beijing Normal University, Zhejiang University, and the Chinese Academy of Science (Zhang, 2020) and supported by the Ministry of Science and Technology (MOST) (Horton et al., 2023). Drs. Long Cao and John Moore made substantial contributions to the scientific literature (see, for instance, Caldeira et al. (2013)).

Though not directly related to SRM, China has seen a number of projects related to weather modification (Chien et al., 2017; Qiu & Cressey, 2008), the most famous of which is weather modification during the 2008 Olympic Games, during which the spectators saw a very blue open sky during the opening ceremony. Weather modification (fog and rain suppression, artificial rain- and snow-making, etc.), is an important policy area and institutionalized (Bluemling et al., 2020), and operates on a daily basis across provinces. China has also invested in large-scale projects such as extensive afforestation and reforestation, the South-North Water Diversion, the Three Gorges Dam (Moore et al., 2016), and the Sky River project that aims to modify the transport of atmospheric moisture through weather modification (Yeh, 2022). Given this history of large-scale engineering projects, some might be tempted to conclude that China is likely to be seriously engaged in SRM research, and potentially, development.

Furthermore, as an emerging economy with increasing scientific capabilities, China is engaged in various scientific projects such as the space station and the international fusion project. While China is developing its own space station, Tiangong, it is a member country of the International Thermonuclear Experimental Reactor (ITER).

Nonetheless, the literature shows that China is not likely to initiate SRM. Dai et al. (2021) conducted an expert survey with Chinese and American scholars on solar geoengineering, and found similar attitudes across the two countries. Edney and Symons (2014) also examined the potential role of China in SRM, and concluded that China is likely to be cooperative in a multi-national framework on SRM. Moore et al. (2016) also counter this argument, drawing on of philosophical traditions and the analysis of the current political landscape.

Unlike experts, Chinese citizens may be more open to the idea of SRM. Visschers et al. (2017) presented a survey with representative samples in China and five Western countries: Canada, Germany, Switzerland, the UK and the USA. Chinese respondents expressed a greater level of support for SRM than Western counterparts. Similarly, a survey study with undergraduate students in six Asia-Pacific countries found that the respondents from China, India, and the Philippines are more supportive of SRM than those from Australia, Japan, and South Korea (Sugiyama et al., 2020).

Lastly, it is important to note philosophical traditions in China. First, the conception of nature could be different (Moore et al., 2016). Second, how the public should be consulted can be affected by the philosophical tradition. As Wong (2013) argued, the public respect the decisions by virtuous and capable leaders. In such a context, the role of public engagement could be different.

4 Country Perspective: India

As with China, India is proactive in terms of both science and governance. India has contributed to natural science such as GeoMIP as well as governance research. On the one hand, Prof. Bala from the Indian Institute of Science, and Dr. Arunabha Ghosh of the Council on Energy, Environment, and Water (CEEW), have been active in science and governance, respectively (Bala & Gupta, 2019; Bala et al., 2019, 2023; Ghosh, 2018; Reynolds et al., 2022). CEEW has been organizing international conferences on governance since 2011 (Bala et al., 2019). The Indian experts are well connected to the international network through various channels such as the IPCC (Intergovernmental Panel On Climate Change, 2023) and the US think tank, Resources for the Future (RFF) (Bala et al., 2023), and C2G (Reynolds et al., 2022) and numerous scientific works. The Department of Science and Technology has been supporting geoengineering research since 2012 (Bala et al., 2019). Nevertheless, such efforts are rather limited compared to other climate-related issues.

The role of India is particularly important not only because it is a major emitter of greenhouse gases and a large, populous country, but also because SRM might affect the Asian monsoon (Asutosh et al., 2025), which is critical for providing precipitation to the climate-dependent agriculture sector of India. Bala and Gupta (2019) said monsoon and climate are synonymous. The possible disruption of the Asian monsoon or more generally precipitation patterns due to SRM has been a long standing concern (Krishnamohan & Bala, 2022; Robock et al., 2009; Tilmes et al., 2013), though some newer studies suggest a potential way around this grave risk (Irvine et al., 2019).

Recently, Bala et al. (2023) conducted a survey with experts on solar geoengineering research, and confirmed that little research has been done in India to date. While respondents support the strengthening of research in India with public funding, they also broadly support many of the governance principles, including transparency and public disclosure.

In terms of public perceptions, the literature is scarce. Sugiyama et al. (2020) confirmed that Indian undergraduate students, like their counterparts in China and the Philippines, were more supportive of solar geoengineering research than those from the surveyed OECD countries.

Just like China, as an emerging economy, India has been engaged with large-scale science projects such as space exploration. India successfully completed a moon landing of a spacecraft near the south pole ("India's Moon Landing Is a Stellar Achievement—And a Win for Science", 2023). India is also involved in the ITER fusion research project. India has also proposed a large-scale engineering project to interlink rivers for irrigation, although its merit is disputed (Chauhan et al., 2023).

5 Country Perspective: Japan

Though Japan is part of the West from the viewpoint of current international affairs, it sometimes differs in terms of some policy issues. Also, its activities are not always fully documented in English, and it is useful to have a close look at it.

There have been some research activities in Japan concerning SRM from both natural science and social science perspectives. For example, scientists from Japan have been actively contributing to GeoMIP (Geoengineering Model Intercomparison Project) (Visioni et al., 2023), part of the broader CMIP (Shingo Watanabe and colleagues led MIROC-ESM analysis of SRM) (Watanabe, 2023). In fact, a small portion of large climate modeling projects has been dedicated to solar geoengineering (as part of S-10, SOUSEI, TOUGOU, programs funded by the Ministry of Education).

More recently, as part of a large-scale, cabinet-level, mission-oriented programs, the Government of Japan began a Moonshot program on weather and typhoon modification (JST, 2023). One of the awardees of the program is the Typhoon Science and Technology Research Center at Yokohama National University, which touts "typhoon innovation" (Fudeyasu, n.d.). In fact, Japan has been engaged with artificial precipitation for a long time. In 2013, the Tokyo Prefecture Government ran an artificial rainmaking device to counteract the prolonged dry weather (Japan Times, 2013). These projects are not directly related to SRM or climate intervention, but as noted above, the mere existence of this project implies the interest of Japan in this kind of project.

Despite a few research initiatives, the discussion on SRM is quite limited. The author of this chapter wrote two popular books on solar geoengineering to facilitate discussion on SRM (not just SRM but also broader climate change as the concern about climate change is lower compared to that of other countries) and led a project to co-design a transdisciplinary research agenda on stratospheric aerosol injection, the most oft-discussed option of SRM (Sugiyama et al., 2017a, 2017b). We do not see lively discussion about SRM, either from pro or con perspectives. There are only few researchers who have expressed concerns about SRM either (except for Asayama & Hulme, 2019 and Kuwata, 2017, 2018). A few expert surveys have been conducted (Lau, 2019; Takahashi et al., 2014) Such studies also confirm the point made above.

On the policy front, it is unlikely for Japan to take leadership in the research and development of solar geoengineering. First, the country doesn't have many resources because of a stagnant economy and an aging population. Second, "extreme options" do not enjoy much support in the country, perhaps because of the smaller number of people and researchers concerned about climate change (Sugiyama, 2020).

Nevertheless, Japan has contributed to a number of large-scale research and development programs across the research areas, including space exploration, nuclear fusion. Japan is unlikely to be in a leading position but could contribute to an international research program.

6 Discussion and Conclusions

This chapter provided a brief review of global discussions on SRM, which was followed by a more in-depth look at three Asian countries: China, India, and Japan.

Survey results show that the awareness of SRM globally is low. They also reveal that those from the Global South tend to be more open to the notion of SRM than the counterparts in the Global North, perhaps because of potentially large damages from climate change on their soil. Nevertheless, they are equally cautious. Public discussions are often dominated by experts from North America and Western Europe, though we see increasing participation from the Global South thanks to initiatives such as the Degrees Initiative and the Alliance for Just Deliberation on Solar Geoengineering.

In spite of different cultural backgrounds and conceptions of nature, the overall attitudes of the citizens and experts are broadly similar to those in North America and Western Europe, except that developing countries may be more open to SRM. A review of research activities and expert and public perception studies demonstrate that research activities in China, India, and Japan are limited, and suggest that the need for further research is appreciated in all the three countries. However, the attitude of experts toward SRM deployment is as cautious as in developed countries in North America and Western Europe. The need for international collaboration is also acknowledged. As with these countries, we see opposing views too; there are some signatories from these countries of the open letter calling for a non-use agreement of solar geoengineering (https://www.solargeoeng.org/non-use-agreement/signatories/). The debate may not be as lively as in Europe or North America, but the general pattern seems to be fairly similar to that of Europe and North America.

Nevertheless, the current situation might be simply a result of the present condition being at an early stage. As science advances and international negotiations proceed, states might develop their positions, and we could see divergence between Asian countries and North America and Western Europe.

Moving forward, in the (second-best) ideal world (the first best being mitigation moving rapidly), the international community should pursue research and development of SRM in an international, collaborative project (there are potentially bad scenarios (Lockley et al., 2022)). National Academies of Sciences, Engineering, and Medicine (2021) made a very explicit call for international collaboration, and so did other publications (Ghosh, 2018; Sugiyama et al., 2017a, 2017b). These studies outlined some possible avenues and lessons. How to foster such collaboration in a concrete manner constitutes an important area of research.

References

Abiodun, B., Abdoulaye, B., Ajayi, V. O., Atanas, D., Bonou, F. K., Coulibaly, A., Diedhiou, A., Egbebiyi, T. S., Fotso-Nguemo, T. C., Kamogne, F. D., Klutse, N. A. B., Lennard, C., Nghonda, J. P., Nkrumah, F., Odoulami, R. C., Opijah, F., Quagraine, K. A., & Quenum, G. M. L. D. (2023). Response to opinion, "my continent is not your giant climate laboratory." *DEGREES Initiative.* https://www.degrees.ngo/dmf-scientists-respond-to-nyt-opinion-piece-on-srm/ (accessed December 15, 2024).

African Academy of Sciences, & Solar Radiation Management Governance Initiative. (2013). *Governance of research on solar geoengineering: African perspectives.* Consolidated report of three workshops in Senegal, South Africa, and Ethiopia. African Academy of Sciences (AAS).

Asayama, S., & Hulme, M. (2019). Engineering climate debt: Temperature overshoot and peak-shaving as risky subprime mortgage lending. *Climate Policy, 19*(8), 937–946. https://doi.org/10.1080/14693062.2019.1623165

Asutosh, A., Tilmes, S., Bednarz, E. M., & Fadnavis, S. (2025). South Asian Summer Monsoon under stratospheric aerosol intervention. *NPJ Climate and Atmospheric Science, 8*(1), 3. https://doi.org/10.1038/s41612-024-00875-z

Bala, G., & Gupta, A. (2019). Solar geoengineering research in India. *Bulletin of the American Meteorological Society, 100*(1), 23–28. https://doi.org/10.1175/BAMS-D-18-0122.1

Bala, G., Krishnamohan, K. S., & Gupta, A. (2019). Geoengineering research activities in India. *Journal of Governance, 18*, 299–307.

Bala, G., Sushma, B. S., Murthy, I. K., & Ravindranath, N. H. (2023). *Strategy for promoting interdisciplinary solar geoengineering research in India* (No. Report 23–38). Resources for the Future.

Baum, C. M., Fritz, L., Low, S., & Sovacool, B. K. (2024). Public perceptions and support of climate intervention technologies across the Global North and Global South. *Nature Communications, 15*(1), 2060. https://doi.org/10.1038/s41467-024-46341-5

Beyerl, K., & Maas, A. (2014). *Perspectives on climate engineering from Pacific small island states* (IASS Working Paper). Institute for Advanced Sustainability Studies (IASS).

Biermann, F., & Gupta, A. (2024). A paradigm shift? African countries call for the non-use of solar geoengineering at UN Environment Assembly. *PLOS Climate, 3*(5), Article e0000413. https://doi.org/10.1371/journal.pclm.0000413

Bluemling, B., Kim, R. E., & Biermann, F. (2020). Seeding the clouds to reach the sky: Will China's weather modification practices support the legitimization of climate engineering? *Ambio, 49*(1), 365–373. https://doi.org/10.1007/s13280-019-01180-3

Brutschin, E., Baum, C. M., Fritz, L., Low, S., Sovacool, B. K., & Riahi, K. (2024). Drivers and attitudes of public support for technological solutions to climate change in 30 countries. *Environmental Research Letters, 19*(11), Article 114098. https://doi.org/10.1088/1748-9326/ad7c67

Caldeira, K., Bala, G., & Cao, L. (2013). The science of geoengineering. *Annual Review of Earth and Planetary Sciences, 41*(1), 231–256. https://doi.org/10.1146/annurev-earth-042711-105548

Carnegie Climate Governance Initiative (C2G). (2023). *Global status of activities relating to Solar Radiation Modification and its governance: Ninth Edition.* https://www.c2g2.net/wp-content/uploads/20230914-C2GSRMGlobalStatusUpdate.pdf (accessed December 15, 2024).

Carr, W. A., Preston, C. J., Yung, L., Szerszynski, B., Keith, D. W., & Mercer, A. M. (2013). Public engagement on solar radiation management and why it needs to happen now. *Climatic Change, 121*(3), 567–577. https://doi.org/10.1007/s10584-013-0763-y

Carr, W. A., & Yung, L. (2018). Perceptions of climate engineering in the South Pacific, Sub-Saharan Africa, and North American Arctic. *Climatic Change, 147*(1), 119–132. https://doi.org/10.1007/s10584-018-2138-x

Chauhan, T., Devanand, A., Roxy, M. K., Ashok, K., & Ghosh, S. (2023). River interlinking alters land-atmosphere feedback and changes the Indian summer monsoon. *Nature Communications, 14*(1), 5928. https://doi.org/10.1038/s41467-023-41668-x

Chien, S.-S., Hong, D.-L., & Lin, P.-H. (2017). Ideological and volume politics behind cloud water resource governance—Weather modification in China. *Geoforum, 85*, 225–233. https://doi.org/10.1016/j.geoforum.2017.08.003

Chirapant, S. (2022, November 29). *The king and the cloud—The story behind the Royal Rain-making Project* [2020/12/17]. Royal Thai Consulate-General, Fukuoka. https://fukuoka.thaiembassy.org/en/content/the-king-and-the-cloud-the-story-behind-the-royal?cate=5ee71d3eb58aa61433490366 (accessed December 15, 2024).

Crutzen, P. J. (2006). Albedo enhancement by stratospheric sulfur injections: A contribution to resolve a policy dilemma? *Climatic Change, 77*(3–4), 211–220. https://doi.org/10.1007/s10584-006-9101-y

Dai, Z., Burns, E. T., Irvine, P. J., Tingley, D. H., Xu, J., & Keith, D. W. (2021). Elicitation of US and Chinese expert judgments show consistent views on solar geoengineering. *Humanities and Social Sciences Communications, 8*(1), 18. https://doi.org/10.1057/s41599-020-00694-6

Dannenberg, A., & Zitzelsberger, S. (2019). Climate experts' views on geoengineering depend on their beliefs about climate change impacts. *Nature Climate Change, 9*(10), 769–775. https://doi.org/10.1038/s41558-019-0564-z

Degrees Initiative. (2023). *SRM engagement workshop in Mexico City*. https://www.degrees.ngo/events/mexico-april-2023/ (accessed December 15, 2024).

Delina, L. L. (2020). Potentials and critiques of building a Southeast Asian interdisciplinary knowledge community on critical geoengineering studies. *Climatic Change, 163*(2), 973–987. https://doi.org/10.1007/s10584-020-02921-0

Edney, K., & Symons, J. (2014). China and the blunt temptations of geo-engineering: The role of solar radiation management in China's strategic response to climate change. *The Pacific Review, 27*(3), 307–332. https://doi.org/10.1080/09512748.2013.807865

ETC Group. (n.d.). *Hands off mother earth: Campaign against geoengineering*. https://www.etcgroup.org/content/hands-mother-earth-0 (accessed December 15, 2024).

ETC Group. (2010). *Geopiracy: The case against geoengineering*. ETC Group.

Fiorino, D. J. (1990). Citizen participation and environmental risk: A survey of institutional mechanisms. *Science, Technology, & Human Values, 15*(2), 226–243. https://doi.org/10.1177/016224399001500204

Fleming, J. R. (2010). *Fixing the sky: The checkered history of weather and climate control*. Columbia University Press.

Fudeyasu, H. (n.d.). *Introduction: A message from the director*. Institute for Multidisciplinary Sciences, Typhoon Science and Technology Research Center (TRC), Yokohama National University. https://trc.ynu.ac.jp/en_Introduction.html (accessed December 15, 2024).

Ghosh, A. (2018). Environmental institutions, international research programmes, and lessons for geoengineering research. In *Geoengineering our climate?* (pp. 199–213). Routledge.

High Representative of the Union for Foreign Affairs and Security Policy. (2023). *Joint communication to the European parliament and the council. A new outlook on the climate and security nexus: Addressing the impact of climate change and environmental degradation on peace, security and defence*. European Commission. https://www.eeas.europa.eu/sites/default/files/documents/2023/JOIN_2023_19_1_EN_ACT_part1_v7.pdf (accessed December 15, 2024).

Horton, J. B., Brent, K., Dai, Z., Felgenhauer, T., Geden, O., McDonald, J., McGee, J., Schenuit, F., & Xu, J. (2023). Solar geoengineering research programs on national agendas: A comparative analysis of Germany, China, Australia, and the United States. *Climatic Change, 176*(4), 37. https://doi.org/10.1007/s10584-023-03516-1

Hussain, A., Sipra, H. F. K., Waheed, A., & Ukhurebor, K. E. (2023). *Exploring the stakeholder's perceptions of solar geoengineering in developing countries (Nos. 23–35)*. Resources for the Future (RFF). https://media.rff.org/documents/WP_23-35.pdf (accessed December 15, 2024).

India's Moon Landing Is a Stellar Achievement—And a Win for Science. (2023). *Nature, 620*(7976), 921. https://doi.org/10.1038/d41586-023-02685-4

Intergovernmental Panel On Climate Change. (2023). *Climate Change 2021—The physical science basis: Working group I contribution to the sixth assessment report of the intergovernmental panel on climate change* (1st ed.). Cambridge University Press. https://doi.org/10.1017/9781009157896

IPCC. (2022). Summary for policymakers. In H.-O. Pörtner, D. C. Roberts, E. S. Poloczanska, K. Mintenbeck, M. Tignor, A. Alegría, M. Craig, S. Langsdorf, S. Löschke, V. Möller, & A. Okem (Eds.), *Climate change 2022: Impacts, adaptation and vulnerability*. Contribution of working group II to the sixth assessment report of the intergovernmental panel on climate change (pp. 3–33). Cambridge University Press.

Irvine, P., Emanuel, K., He, J., Horowitz, L. W., Vecchi, G., & Keith, D. (2019). Halving warming with idealized solar geoengineering moderates key climate hazards. *Nature Climate Change, 9*(4), 295–299. https://doi.org/10.1038/s41558-019-0398-8

Japan Times. (2013, August 22). *Cloud-seeding brings rain to Tama*. https://www.japantimes.co.jp/news/2013/08/22/national/cloud-seeding-brings-rain-to-tama/ (accessed December 15, 2024).

JST. (2023). *Moonshot R&D program overview*. Department of Moonshot R&D Program at the Japan Science and Technology (JST) Agency.

Krishnamohan, K. S., & Bala, G. (2022). Sensitivity of tropical monsoon precipitation to the latitude of stratospheric aerosol injections. *Climate Dynamics, 59*(1–2), 151–168. https://doi.org/10.1007/s00382-021-06121-z

Kuwata, M. (2017). Jinshinsei to kikō kōgaku [The Anthropocene and climate engineering].*Gendai Shisō [Review of Contemporary Thought], 45*(22), 122–131. (In Japanese)

Kuwata, M. (2018). Kikō kōgaku to katasutorofi [Climate engineering and catastrophe]. In A. Yoshinaga & M. Fukunaga (Eds.),*Mirai no kankyō rinrigaku [Future environmental ethics]* (pp. 125–140). Keisō Shobō. (In Japanese) https://cir.nii.ac.jp/crid/1010000781827567494

Lau, S. (2019). *Geoengineering the climate: A snapshot of Japanese expert opinions*. University of Tokyo.

Lockley, A., Xu, Y., Tilmes, S., Sugiyama, M., Rothman, D., & Hindes, A. (2022). 18 politically relevant solar geoengineering scenarios. *Socio-Environmental Systems Modelling, 4*, 18127. https://doi.org/10.18174/sesmo.18127

McLaren, D., & Corry, O. (2025). Solar geoengineering research faces geopolitical deadlock. *Science, 387*(6729), 28–30. https://doi.org/10.1126/science.adr9237

Moore, J. C., Chen, Y., Cui, X., Yuan, W., Dong, W., Gao, Y., & Shi, P. (2016). Will China be the first to initiate climate engineering? *Earth's Future, 4*(12), 588–595. https://doi.org/10.1002/2016EF000402

Munoz, L. (2017). *Seeding change in weather modification globally*. World Meteorological Organization. Retrieved from https://public.wmo.int/en/resources/bulletin/seeding-change-weather-modification-globally (accessed December 15, 2024).

National Academies of Sciences, Engineering, and Medicine. (2021). *Reflecting sunlight: Recommendations for solar geoengineering research and research governance* (p. 25762). National Academies Press. https://doi.org/10.17226/25762

Okereke, C. (2023, April 18). My continent is not your giant climate laboratory. *New York Times*. Retrieved from https://www.nytimes.com/2023/04/18/opinion/geoengineering-climate-change-technology-africa.html (accessed December 15, 2024).

Pasztor, J. (2021). Solar geoengineering research needs formal global debate. *Nature, 595*(7868), 494–494. https://doi.org/10.1038/d41586-021-01957-1

Pasztor, J., Scharf, C., & Schmidt, K.-U. (2017). How to govern geoengineering? *Science, 357*(6348), 231–231. https://doi.org/10.1126/science.aan6794

Qiu, J., & Cressey, D. (2008). Meteorology: Taming the sky. *Nature, 453*(7198), 970–974. https://doi.org/10.1038/453970a

Rahman, A. A., Artaxo, P., Asrat, A., & Parker, A. (2018). Developing countries must lead on solar geoengineering research. *Nature, 556*(7699), 22–24. https://doi.org/10.1038/d41586-018-03917-8

Reynolds, J. (2019). *The governance of solar geoengineering: Managing climate change in the Anthropocene*. Cambridge University Press.

Reynolds, J., Ghosh, A., Harihar, N., & Jain, P. (2022). *Solar radiation modification: Governance gaps and challenges*. Carnegie Climate Governance Initiative (C2G).

Ricke, K. L., Moreno-Cruz, J. B., & Caldeira, K. (2013). Strategic incentives for climate geoengineering coalitions to exclude broad participation. *Environmental Research Letters, 8*(1), Article 014021. https://doi.org/10.1088/1748-9326/8/1/014021

Robock, A., Marquardt, A., Kravitz, B., & Stenchikov, G. (2009). Benefits, risks, and costs of stratospheric geoengineering. *Geophysical Research Letters, 36*(19), L19703. https://doi.org/10.1029/2009GL039209

Stirling, A. (2008). Opening up" and "closing down": Power, participation, and pluralism in the social appraisal of technology. *Science, Technology, & Human Values, 33*(2), 262–294. https://doi.org/10.1177/0162243907311265

Sugiyama, M. (2020). Kikō kōgaku (jio-enjiniaringu) ni kakawaru senmonka to sono komyunitī: Nihon to Ōbei no chigai no ichikōsatsu [Climate engineering (geoengineering) experts and their community: A consideration of the differences between Japan and the West]. *Kankyō Shakaigaku Kenkyū [Journal of Environmental Sociology], 26*, 24–43. (In Japanese) https://doi.org/10.24779/jpkankyo.26.0_24

Sugiyama, M., Asayama, S., Ishii, A., Kosugi, T., Moore, J. C., Lin, J., Lefale, P. F., Burns, W., Fujiwara, M., Ghosh, A., Horton, J., Kurosawa, A., Parker, A., Thompson, M., Wong, P.-H., & Xia, L. (2017a). The Asia-Pacific's role in the emerging solar geoengineering debate. *Climatic Change, 143*(1), 1–12. https://doi.org/10.1007/s10584-017-1994-0

Sugiyama, M., Asayama, S., & Kosugi, T. (2020). The North-South divide on public perceptions of stratospheric aerosol geoengineering?: A survey in six Asia-Pacific Countries. *Environmental Communication, 14*(5), 641–656. https://doi.org/10.1080/17524032.2019.1699137

Sugiyama, M., Asayama, S., Kosugi, T., Ishii, A., Emori, S., Adachi, J., Akimoto, K., Fujiwara, M., Hasegawa, T., Hibi, Y., Hirata, K., Ishii, T., Kaburagi, T., Kita, Y., Kobayashi, S., Kurosawa, A., Kuwata, M., Masuda, K., Mitsui, M., … Yoshizawa, G. (2017b). Transdisciplinary co-design of scientific research agendas: 40 research questions for socially relevant climate engineering research. *Sustainability Science, 12*(1), 31–44. https://doi.org/10.1007/s11625-016-0376-2

Sugiyama, M., Asayama, S., Kosugi, T., Ishii, A., & Watanabe, S. (2025). Public attitude toward solar radiation modification: Results of a two-scenario online survey on perception in four Asia-Pacific countries. *Sustainability Science, 20*(2), 423–438. https://doi.org/10.1007/s11625-024-01520-7

Takahashi, Y., Hirose, A., Kawamura, R., Kaneko, H., & Sugiyama, M. (2014). Kikō kōgaku (jio-enjiniaringu) no tekunorojī asesumento ni shisuru senmonka e no kikitori chōsa to sono bunseki: Nihon ni okeru ichikentō [Interview survey of experts contributing to technology assessment of climate engineering (geoengineering) and its analysis: A study in Japan]. *Kenkyū Gijutsu Keikaku [The Journal of Science Policy and Research Management], 29*(2/3), 214–228. (In Japanese). https://doi.org/10.20801/jsrpim.29.2_3_214

Temple, J. (2023, April 17). This technology could alter the entire planet. These groups want every nation to have a say. *MIT Technology Review*. https://www.technologyreview.com/2023/04/17/1071654/this-technology-could-alter-the-entire-planet-these-groups-want-every-nation-to-have-a-say/ (accessed December 15, 2024).

Tilmes, S., Fasullo, J., Lamarque, J., Marsh, D. R., Mills, M., Alterskjær, K., Muri, H., Kristjánsson, J. E., Boucher, O., Schulz, M., Cole, J. N. S., Curry, C. L., Jones, A., Haywood, J., Irvine, P. J., Ji, D., Moore, J. C., Karam, D. B., Kravitz, B., … Watanabe, S. (2013). The hydrological impact of geoengineering in the Geoengineering Model Intercomparison Project (GeoMIP). *Journal of Geophysical Research: Atmospheres, 118*(19). https://doi.org/10.1002/jgrd.50868

Visioni, D., Kravitz, B., Robock, A., Tilmes, S., Haywood, J., Boucher, O., Lawrence, M., Irvine, P., Niemeier, U., Xia, L., Chiodo, G., Lennard, C., Watanabe, S., Moore, J. C., & Muri, H. (2023). Opinion: The scientific and community-building roles of the Geoengineering Model Intercomparison Project (GeoMIP)—Past, present, and future. *Atmospheric Chemistry and Physics, 23*(9), 5149–5176. https://doi.org/10.5194/acp-23-5149-2023

Visschers, V. H. M., Shi, J., Siegrist, M., & Arvai, J. (2017). Beliefs and values explain international differences in perception of solar radiation management: Insights from a cross-country survey. *Climatic Change, 142*(3), 531–544. https://doi.org/10.1007/s10584-017-1970-8

Watanabe, S. (2023). Taiyo housha kaihen no kagakuteki rikai no genjo to kadai [Current status and issues in scientific understanding of solar radiation modification]. *Enerugi Shigen* [*Energy and Resources*], *44*(3), 166–170.

White House Office of Science and Technology Policy (OSTP). (2023). *Congressionally mandated research plan and an initial research governance framework related to solar radiation modification.* https://www.whitehouse.gov/ostp/news-updates/2023/06/30/congressionally-mandated-report-on-solar-radiation-modification/ (accessed December 15, 2024).

Winickoff, D. E., Flegal, J. A., & Asrat, A. (2015). Engaging the Global South on climate engineering research. *Nature Climate Change, 5*(7), Article 7. https://doi.org/10.1038/nclimate2632

Wong, P.-H. (2013). The public and geoengineering decision-making: A view from Confucian political philosophy. *Techné: Research in Philosophy and Technology, 17*(3), 350–367. https://doi.org/10.5840/techne201421110

Yeh, E. T. (2022). Sky river: Promethean dreams of optimising the atmosphere. *Made in China Journal, 7*(2), 96–101. https://doi.org/10.22459/MIC.07.02.2022.09

Zhang, Z. (2020, November 9). Has 'geoengineering' arrived in China? *China Dialogue.* https://chinadialogue.net/en/climate/how-to-supervise-geoengineering/ (accessed December 15, 2024).

Who Will Bell the Cat? African Union, Geoengineering and the Potential Role of Human Rights

Ademola Oluborode Jegede

Abstract There is increasing scholarship on the necessity for geoengineering to counteract climate change. In the discourse, carbon dioxide removal (CDR) options have featured prominently. Despite its promises, bourgeoning scholarship shows that CDR options are not without uncertainties that may have implications for a wide range of human rights in developing countries. The African Union (AU) human rights standards are anchored in the African Charter on Human and Peoples' Rights (African Charter), a pillar instrument monitored by the African Commission on Human and Peoples' Rights (African Commission) and complemented by the African Court of Human and Peoples' Rights. The African Charter is supported by other policy and legal instruments that reflect the commitment of states to addressing climate change. However, there is rare research on the trend in geoengineering, especially in CDR, as well as the adequacy or shortcomings of its policy and legal framework at the AU level. Consequently, this paper interrogates the following issues: the trends in geoengineering, with a focus on the positive and negative implications of carbon removal measures within the AU; the potential of the AU's normative framework on climate change to support climate engineering; and the ways in which state human rights obligations under the African Charter may shape key AU legal and policy provisions to minimize the anticipated risks while implementing CDR geoengineering.

Keywords African Union · CDR · Geoengineering options · Human rights · State obligation

---

A. O. Jegede (✉)
Ismail Mahomed Centre for Human and Peoples' Rights, University of Venda, Thohoyandou, South Africa
e-mail: ademola.jegede@univen.ac.za

K.-U. Schrogl et al. (eds.), *The Politics of Geoengineering*, Springer Climate,
https://doi.org/10.1007/978-3-032-10095-5_6

1 Introduction

Geoengineering refers to 'the deliberate large-scale manipulation of the planetary environment to counteract anthropogenic climate change', as a climate intervention option (The Royal Society, 2009, p. 15). It entails various strategies to modify the Earth's natural systems, such as the atmosphere, soil, and oceans, in an attempt to address human-induced (anthropogenic) climate change (Bellamy et al., 2017). While it has been a subject of debate for long (Keane, 2020, p. 56), recent attention has been drawn to the idea due to the reality that it is nearly impossible for conventional climate response measures alone to achieve net zero carbon dioxide (CO_2) emissions and stabilize global temperatures below 2 °C (Stocker et al., 2013, p. 25; The Royal Society and Royal Academy of Engineering, 2018, p. 7). Hence, the purpose of geoengineering is to cut down on the levels of greenhouse gas (GHG) emissions globally (Marshall, 2017, p. 183). Options for climate geoengineering include solar radiation management (Keane, 2020, p. 56), marine cloud brightening (MCB) (Geoengineering Technology Briefing, 2021), and CDR techniques (Hale & Dilling, 2011, p. 190). Of these strategies, CDR, which is the focus of this paper, is the most discussed in the agenda of climate policy (Morrow et al., 2020, pp. 150–153). The Intergovernmental Panel on Climate Change (IPCC) Sixth Assessment Report (AR6) recognizes that the CDR options may play a role in achieving global ambition (IPCC, 2023, pp. 1–34).

Implementing CDR options touches on a whole range of resources such as land, water, food, and energy that can affect human welfare and national development in developing countries (Smith et al., 2016, p. 8). Bourgeoning scholarship and reports provide that the implications of CDR have certain and clear impacts on livelihood and human rights (Jegede, 2021). This development makes the human rights field important to the governance of CDR. A human rights framework, affirms Burns (2016, pp. 150–170), is essential in addressing anticipated justice issues around the deployment of CDR. Other writings affirm that a human rights framework is crucial as it offers a normative framework and language of governance for climate geoengineering interventions at both international and national levels (Jegede, 2021; Svoboda et al., 2019, p. 397).

Africa contributes approximately 3–4% of global carbon emissions (Al-Zu'bi et al., 2022, pp. 1078–1084), however, it is not exempted from the new order of global commitment under the Paris Convention to reduce carbon dioxide emissions and limit global warming to 1.5 °C or 2 °C (Paris Agreement, 2015). The continent is committed to undertaking efforts to reduce carbon dioxide emissions (Armani et al., 2022). Geoengineering, especially CDR, is of significance to Africa due to its natural ecosystem. In fact, the continent is the second largest landmass globally, with a total land area of approximately 30.37 million square kilometers and holds 65% of the world's arable land (AFDB, 2018). Also, Africa is highly diverse, with a unique ecosystem that comprises of mangroves, forests, woodlands, savannas, deserts, grasslands, arid lands, and rangelands (FAO & AUDA-NEPAD,

2021). While the involvement of the AU is crucial to the development and formulation of appropriate and standalone policy on the implementation of CDR, this is not yet the case.

However, a range of existing instruments at the AU level may hold significant relevance for advancing CDR initiatives. At that level, there is a human rights system principally anchored in the African Charter on Human and Peoples' Rights (OAU, 1981). The African Charter is a pillar human rights instrument, having been ratified by 54 member states of the AU (African Commission on Human and Peoples' Rights, n.d.). In particular, the human rights provisions of the African Charter are principally monitored by the African Commission on Human and Peoples' Rights (African Commission) which has rich jurisprudence and applies international law by virtue of article 61 of the African Charter. Other policy and legal frameworks of the AU that are significant to addressing climate change are evident in instruments such as the African Union Convention for the Protection and Assistance of Internally Displaced Persons in Africa (African Union, 2009), the African Convention on the Conservation of Nature and Natural Resources (African Union, 2003), the African Leaders Nairobi Declaration on Climate Change and a Call to an Action (African Heads of State & Government, 2023), African Union Climate Change and Resilient Development Strategy and Action plan 2022–2032, Framework and Reform Strategy for Fisheries and Aquaculture in Africa, the African Union Blue Economy Strategies, and Agenda 2063. However, the utility of the foregoing instruments in the context of coordinating the implementation of the CDR geoengineering at the AU level is hardly examined and has not been developed.

2 AU and CDR: Opportunities and Challenges

The participation of the African continent in geoengineering projects and discussions has been limited. As Chalmin (2021) reports, only 5% of the total documented geoengineering projects were executed on the African continent, and these were externally driven. Additionally, the participation of organs of the AU in the discussion on geoengineering is relatively unknown and unproven. For instance, while over the years the AU has consistently maintained a common position in the discussion and negotiation on climate issues under the auspices of the UNFCCC (African Union, 2008, 2014), there is no specific evidence of its official position on geoengineering. Yet Africa is endowed with resources which may be impacted by the agenda of geoengineering, particularly by the CDR options. With a focus on afforestation, bioenergy with carbon capture and storage (BECCS), and ocean fertilization, the discussion in this section demonstrates that there are both opportunities and challenges in implementing Africa-wide CDR options.

Afforestation entails planting or replanting forests over large areas for the purpose of absorbing carbon in both the trees and the soil as they grow (IUCN, 2004). As defined by the UNFCCC, afforestation is a human-induced transformation of non-forested land to forested land through planting, seeding, and/or promotion of natural seed sources. In Africa, this could be helpful in terms of ecosystems and human services (IUCN, 2004, p. 7), and could mitigate environmental degradation and serve as a sink to carbon (Nowak et al., 2007, p. 220). However, afforestation carries potential risks, as numerous studies affirm that forest plantations consume more water than baseline vegetation, leading to a reduction of water yield (streamflow) (Gush, 2006). In fact, afforestation utilizes water that would otherwise form catchment runoff (Turpie et al., 2007, p. 423). Furthermore, afforestation requires large-scale land use which carries a risk for local populations (Malkamaki et al., 2018, p. 90). It can displace customary land title and uses in rural areas whose populations depend on such resources for survival (Cotula et al., 2014, p. 903; Hansungule & Jegede, 2014, p. 256). This signifies that despite its relevance in addressing climate change, afforestation may pose a challenge, in particular to states in Africa that are water stressed. These include states such as: Chad, Comoros, Djibouti, Eritrea, Ethiopia, Liberia, Libya, Madagascar, Niger, Sierra Leone, Somalia, South Sudan, and Sudan (United Nations, 2023).

BECCS involves the growing or collection and the processing of biomass for conversion to heat, electricity, or liquid or gas fuels, capturing the resulting carbon, and storing it underground or in long-lasting products (Burns, 2016; Burns & Nicholson, 2021). This approach has potential for achieving 'net negative' emissions (Thangaraj et al., 2018). In fact, with BECCS, it is likely to capture 90% or more of the carbon dioxide released through biomass production (Koornneef et al., 2012, p. 117). Furthermore, BECCS can help reverse reliance on coal-based energy production, and thereby contribute to global reductions of carbon emissions and the adverse consequences of coal production and burning on populations.

However, BECCS is not without its downsides. The use of food crops such as maize for energy purposes is problematic on a continent that still struggles with food security. Additionally, while Jatropha is a drought-resistant biofuel crop, its use is discouraged because it leaves behind a toxic seedcake (Department of Minerals & Energy, 2007). Generally, major concerns around mass cultivation of food crops for energy purposes remain water limitations, land degradation, and the effects of large-scale production on food security and employment creation (Jonker et al., 2017, p. 1). The foregoing concerns around BECCS reinforce the position of studies which have shown that large-scale deployment of BECCS could have negative implications for food security for some of the world's most vulnerable populations (Burns & Nicholson, 2021). This is because it will require the conversion of substantial areas of land in Africa that are wrongly classified as abandoned and marginal (Smith et al., 2016, p. 42), a development that can result in displacement of the poor from their historical land (Jegede, 2016, pp. 58–82; Kartha & Dooley, 2016). Moreover, its implementation may adversely affect water availability (Smith, 2016, p. 1315) and exacerbate environmental degradation (Kartha & Dooley, 2016, p. 16).

Ocean iron fertilization is a CDR approach by which iron is introduced into iron-poor areas of the ocean surface to trigger phytoplankton production, which can enhance biological productivity and/or accelerate carbon dioxide sequestration from the atmosphere (Lampitt et al., 2008). Proponents of this option contend that phytoplankton production is limited due to low concentrations of iron in the Southern

Ocean, subarctic Pacific, and eastern equatorial Pacific waters (Singh et al., 2015). Therefore, artificially adding iron in these regions could stimulate phytoplankton production and enhance carbon dioxide uptake (Hubbard, 2016, p. 591). However, the significance of ocean fertilization to the carbon removal agenda remains contested. While there are studies which demonstrate that ocean iron fertilization could substantially reduce atmospheric carbon dioxide levels (Eick, 2010, p. 351), other findings conclude that it would result in minimal sequestration of atmospheric carbon and may disrupt marine biodiversity (Ciais et al., 2013).

3 Implications of CDR Geoengineering for Human Rights Under the AU

The AU is yet to release any official, specific, and continental statement on geoengineering. None of the legal and policy instruments relevant to climate change, as will soon be demonstrated, specifically uses the term geoengineering. However, the content of a number of key instruments is not incompatible with CDR. Based on the foregoing, this section argues that implementing CDR options has the dual potential to threaten and enhance key human rights provisions and ambitions of states under the AU. In particular, this includes the right to life, the right to have access to sufficient food and water, the right to have access to health care services, the right to property, the right to adequate housing, and the right to environment, all of which are guaranteed under the Africa Charter and complemented in other AU policy documents. In terms of article 61 of the African Charter, the African Commission may apply international law. Hence, reference is made to UN documents, such as General Comments in discussing these rights.

The right to life is explicitly protected under article 4 of the African Charter, and it is recognized as part of customary international law and the general principles of law (Gormley, 1987, p. 191), and as a jus cogens norm, which is universally binding at all times (Human Rights Committee, 1982). In its General Comment No. 3 of 2015, the African Commission affirmed that the right to life is globally recognized as a foundational human right, and it is of the nature that should not be interpreted narrowly (African Commission on Human and Peoples' Rights, 2015). Rather, for the right to life to be achieved, 'the realisation of all human rights' are a prerequisite (African Commission on Human and Peoples' Rights, 2015). The right to life can be both enhanced and hindered while implementing afforestation programs, for instance. On the one hand, given that trees are central to the human respiratory process and the reduction of environmental degradation (Turner-Skoff & Cavender, 2019, pp. 323–335), they ultimately contribute to the enhancement of the right to life. On the other hand, their heavy reliance on water may undermine the right to life given the centrality of water to life.

Additionally, although not categorically stated in the African Charter, the jurisprudence of the African Commission shows that the right to food is justiciable. In Social and Economic Rights Action Center (SERAC) and Center for Economic and Social Rights (CESR) v Nigeria (SERAC v Nigeria, 2001), the African Commission interpreted articles 4 (right to life), 16 (right to health) and 22 (right to economic, social, and cultural development) to ground a violation of the right to food. While considering the normative content of the right to food, the CESCR General Comment No. 12 of 1999 (CESCR, 1999), urges States parties to note that the normative elements of the right to food, that is, the availability, accessibility, acceptability, and safety of food, can be adversely affected by climatic and ecological factors. It therefore urges States parties to adopt appropriate measures to ensure that climate change does not adversely affect the right to food (CESCR, 1999, pp. 4 and 7). In helping to stabilize the climate system, deploying all the CDR options is, at least indirectly, positive for the right to food in Africa, as it will contribute to reducing extreme events and changes in temperature associated with continued carbon emissions that undermine land suitability and crop yields. However, the diversion of agricultural land for afforestation and BECCS projects can reduce the availability of food, thereby undermining the right to food in states in Africa.

Similarly, the right to water is not specifically mentioned in the African Charter but can be derived from article 16(1). General Comment 15 on the right to water of the CESCR recognizes that the right to water is 'fundamental for life and health' and a 'prerequisite for the realisation of other human rights' (CESCR, 2002). The normative contents of the right to water are availability, quality, and accessibility. Availability requires that the '[w]ater supply for each person must be sufficient and continuous for personal and domestic uses' (CESCR, 2002, para 12(a)); quality entails that water must be safe and free of any substance that is harmful to health (CESCR, 2002, para 12(b)). Therefore, on the one hand, the use of CDR options such as BECCS, ocean fertilization, and afforestation may reverse the adverse effects of climate change such as flooding, which weaken the infrastructure necessary to achieve availability and delivery of clean water to populations, enhancing access to water. On the other hand, due to water demand and its limitation, the options of afforestation and BECCS may threaten the right to water.

More generally, article 16 of the African Charter ensures the right to enjoy the best attainable state of physical and mental health, and that States must ensure that everyone has access to medical care. In 2008, the African Commission adopted a resolution on maternal mortality. In this resolution, the African Commission declared maternal mortality a state of emergency in Africa and called on states to take decisive measures to address this concern (African Commission on Human and Peoples' Rights, 2008). While interpreting the right to health, CESCR General Comment No. 14 identifies determinants of health, including the prevention and reduction of exposure to harmful substances such as radiation, harmful chemicals, or other detrimental environmental conditions that directly or indirectly impact upon human health. Building on this understanding, the implementation of CDR options may stabilize temperature, reduce extreme weather events, prevent the spread of associated epidemics, and thereby contribute to the realization of the right to health. However, the potential impact of certain CDR options, such as ocean fertilization, could release nitrous oxide and other pollutants, increasing pollution and thereby undermining the realization of the right to health.

Article 14 of the African Charter guarantees the right to property, providing that it can be limited only in the interest of public policy and in accordance with the provision of the law. The African Commission has been flexible on what constitutes the concept of 'property'. For instance, in Malawi African Association and Others v Mauritania (ACHPR, 2000, para 128), land was considered 'property' for the purposes of Article 14 of the African Charter. In Centre for Minority Rights Development (Kenya) and Minority Rights Group (on behalf of Endorois Welfare Council) (Centre for Minority Rights Development, 2009, para 187), traditional land was regarded by the African Commission as constituting 'property' under the Charter. For its potential to undermine and dispossess local populations of their customary land tenure system and traditional use of land, the deployment of CDR options of afforestation and BECCS may affect adversely the rights of the population to property and adequate housing. Nonetheless, if implemented with due regard for the nature of land tenure of populations, afforestation and BECCS can accommodate the interest of local populations and boost the realization of property or adequate housing in African states.

The African Charter is silent on the right to adequate housing. To fill this gap, the African Commission stated that the combined effect of articles 14, 16 and 18(1) constitutes a right to shelter or housing. It further asserts that 'the right to adequate housing as implicitly protected in the Charter includes the right to protection against forced evictions' (SERAC v Nigeria, 2001, para 63). General Comment No. 4 of the CESCR enunciates the seven elements constituting the normative content of the right to adequate housing (CESCR, 1991). These are: legal security of tenure, availability of services, materials, facilities and infrastructure, accessibility, location, affordability, habitability, and cultural adequacy (CESCR, 1991). This is further reinforced by the CESCR General Comment 7 on the right to housing, which deals with the issue of forced evictions (CESCR, 1997), defined as: '[p]ermanent or temporary removal against their will of individuals, families and/or communities from the homes and/or land which they occupy, without the provision of, and access to, appropriate forms of legal or other protection' (CESCR, 1997). Therefore, if designated and executed in alignment with these principles, interventions such as afforestation and BECCS could potentially enhance housing conditions and mitigate forced evictions, thereby promoting broader social and economic rights under the African Charter.

Article 24 of the African Charter guarantees that 'all peoples shall have the right to a general satisfactory environment favorable to their development'. This right focuses on the protection and wellbeing of both peoples and their environment. General Comment 14 of the CESCR on the Right to the Highest Attainable Standard of Health (CESCR, 2000), includes a healthy environment as a determinant of health (CESCR, 2000, para 4), and calls for healthy occupational and environmental conditions (CESCR, 2000, para 11). In the Ogoniland case, the African Commission asserts that a generally satisfactory environment connotes 'a clean and safe environment that is closely linked to economic and social rights in so far as the environment affects the quality of life and safety of the individual' (SERAC v Nigeria, 2001). The potential risk of pollution associated with ocean fertilization stands in stark contrast to the important objective of preventing pollution, particularly when it threatens biodiversity or contaminates the sea. However, over the long term, deploying CDR options can offer significant benefits by reducing if not eliminating developmental choices of the state that are harmful to a sustainable environment. For instance, BECCS can constitute an alternative to reliance on coal-based energy systems in Africa and its associated environmental effects, thereby enhancing the realization of the provisions outlined in section 24.

The foregoing potentials are within the scope of a range of policy instruments of the AU which feature climate change and its implications for Africa. For instance, the 2023 African Leaders Nairobi Declaration on Climate Change and a Call to an Action reiterates that 'Africa possesses both the potential and the ambition to be a vital component of the global solution to climate change', and commits to the advancement of 'green industrialization across the Continent by prioritizing energy-intense industries to trigger a virtuous cycle of renewable energy deployment and economic activity' (African Heads of State & Government, 2023, para 5). The African Union Climate Change and Resilient Development Strategy and Action Plan affirms the urgency of building a climate-resilient Africa and identifies the CDR options as part of the innovative and technological approach being proposed to fight climate change (African Union, 2022, p. 28). It also acknowledges that the governance of CDR is an emerging issue on which African government officials, policy bodies, and relevant stakeholders have not been fully engaged (African Union, 2022, p. 32). Yet, there is urgency for Africa to understand the role that CDR could play in African or global climate policy as the continent has great potentials to shape the discussions. The policy further provides that other carbon removal activities such as bio energy and carbon capture and storage, if administered sustainably, could serve as carbon removal from the atmosphere and generate energy to foster financial flow and to meet obligations for the livelihood enabling expansion and development in Africa (African Union, 2022, p. 71).

In alignment with this approach, the African Union Blue Economy Strategies provides that 'blue carbon is part of blue economy with an opportunity to develop coastal wetland to mitigate climate change'. It emphasizes the promotion and integration of blue carbon and ecosystem services into climate change policies and coastal and aquatic policies, while also fostering the development of environmentally sustainable and climate-resilient economies and communities (AU-IBAR, 2019, pp. 8–9). Carbon capture is one of the strategies that is included in the document as a means of realizing its aspirations (AU-IBAR, 2019, p. 10). Also, the Framework and Reform Strategy for Fisheries and Aquaculture in Africa (AUDA-NEPAD, 2014) highlights the importance of regional collaborative management, emphasizes strengthening south-south cooperation, and advocates for the 'African Voice' in international policy dialogues, all of which have significant implications for African fisheries governance. Additionally, the Agenda 2063 commits Africa to climate interventions by focusing on adaptation while relying on multidisciplinary skills, technology transfer, and financial assistance to ensure implementation of interventions (African Union, 2014, p. 16). It further commits Africa to continuing its participation to climate mitigation, supporting sustainable development (African Union,

2014, p. 17). Moreover, the Agenda 2063 underscores the need to establish regional technology centers that collaborate with national designated climate technology entities, develop climate change programs targeting women and youth, and implement climate-resilient agricultural development initiatives (African Union, 2014, p. 72(f)).

Of further significance to CDR options is article 7(1) of the African Convention on the Conservation of Nature and Natural Resources (Revised version), also known as the Conservation Convention 2003, which requires qualitative and quantitative preservation and management of water resources at the highest possible level. The Convention also calls upon states to take measures intended to: '(a) maintain water-based essential ecological procedures as well as to shield human health against pollutants and water-borne diseases; (b) prevent harm that could affect human health or natural resource in another state by the discharge of pollutants; and (c) prevent excessive abstraction, to the profit of downstream communities and states' (African Union, 2003, art. 7(1)).

Finally, the Kampala Convention is important as it not only acknowledges that displacement can arise from 'natural or human-made disasters' but also specifically refers to the obligations of states to protect and assist persons displaced by climate change. In particular, article 5(4) requires states to take measures to protect and assist persons who have been internally displaced due to natural or human-made disasters, including climate change. Additionally, article 4(5) recognizes that those adversely affected by climate change may include indigenous communities and local populations with 'special attachment to, and dependency, on land due to their particular culture and spiritual values'. It obligates parties to ensure that such communities are not displaced from their lands, except for compelling and overriding public interests (African Union, 2009, art. 4(5)). Furthermore, article 10 specifically requires states to prevent displacement caused by projects carried out by private or public actors, emphasizing the need for consultations and environmental impactful assessments prior to undertaking such projects. The application of the foregoing provisions may be useful in addressing the risks associated with the implementation of CDR options, provided State human rights obligations are adequately deployed to minimize the anticipated risks of CDR geoengineering. This forms the focus of the next section.

4 CDR and State Human Rights Obligations Under of CDR

The human rights tripartite typology of duties is: (1) duties to avoid the deprivation of the right concerned, (2) duties to protect rights holders from deprivation, and (3) duties to aid rights holders who have been deprived (Shue, 1980, p. 52). The human rights obligations related to climate change have been explored under the aegis of the United Nations by the Human Rights Council, and the special procedures (Human Rights Council resolutions 7/23, 10/4, 18/22, 26/27, 29/15, 32/33, 35/20, and 38/4; A/HRC/10/61, A/HRC/32/23, A/HRC/35/13, A/HRC/38/21, A/HRC/41/26, A/64/255, A/67/299, A/HRC/36/46, A/HRC/41/39, A/HRC/16/43, A/66/285, A/HRC/31/

52, A/HRC/25/53). Based on its interpretation of the African Charter, the African Commission identifies four layers of obligations applicable to the rights guaranteed under the Charter. In the Ogoniland case, in the context of environmental claims over the degradation of the land of the Ogoni people, the African Commission developed jurisprudence outlining these four obligations—namely, the obligation to 'respect', 'protect', 'promote', and 'fulfil' civil, political, and socio-economic rights.

According to the African Commission, the obligation to respect entails that states should not interfere in the enjoyment of human rights. It signifies that there should be respect on the part of the state for 'right-holders, their freedoms, autonomy, resources, and liberty of their action' (SERAC v Nigeria, 2001, para 45). In the context of CDR, this means that states in Africa should not embark upon actions that pose further threat to the climate. Furthermore, states must put an end to existing climate actions that are harmful to human livelihood and environment. In that regard, states must respect the 'do no harm', the 'preventive principle', and the 'precautionary principle' when making decisions on climate change-related projects or actions. The no-harm principle, as established in Trail Smelter (1941) and confirmed by later international authorities (ICJ, 1996, p. 29), requires that no State has the right to use or permit its territory in a manner that causes serious injury to the territory of another or the properties or persons therein. The preventive principle emphasizes the State's duty to take proactive steps to minimize the risk of damages. As stated in Pulp Mills, international law requires States 'to use all the means at its disposal in order to avoid activities which take place in its territory or any area under its jurisdiction, causing significant damage to the environment of another State' (ICJ, 2010, para. 101).

Finally, the precautionary principle, as reflected in article 3 of the UNFCCC, demands that the lack of full scientific certainty should not be used as a reason for anticipating, preventing or minimizing the causes of climate change and mitigating its adverse effects. It demands that states must not support or approve any act by third parties, especially businesses that threaten the climate system. Also, in line with this obligation, they must avoid discrimination and retrogressive measures in the implementation of CDR. States in Africa must avoid the expropriation of resources of vulnerable groups for climate actions without appropriate consultation and consent of those belonging to such communities, and without adequate compensation. In Front for the Liberation of the State of Cabinda v Republic of Angola (ACHPR, 2013), the African Commission reaffirmed its jurisprudence, emphasizing that the state's general role with respect to property right is 'to respect and protect this right against any form of encroachment, and to regulate the exercise of this right in order for it to be accessible to everyone'. Accordingly, it is important that states avoid expropriation without compensation while implementing CDR options that require land resources.

In discussing the obligation to protect, the African Commission enjoins the state to adopt measures, including legislations, and provide effective remedies in protection of right-holders 'against political, economic and social interferences'. It further requires the regulation of non-state actors to ensure that their operation does not hinder the realization of rights (SERAC v Nigeria, 2001, p. 46). In the Kilwa decision, one of the reasons the African Commission found a violation of article 1 of the African Charter was that the State ‘failed not only to investigate and punish the involvement of the Anvil Mining Company but also to provide redress for the victims against the Company for the role it played in the perpetration of the violations’ (IHRDA and Others v DRC, 2016). Building on this, the obligation to protect requires positive acts by states to enact and enforce climate related laws that safeguard the rights of populations affected by climate change. This implies that states must establish an important legal framework to address the diverse aspects of climate geoengineering in Africa. States’ obligation to protect individuals and groups extends to foreseeable threats to human rights arising from CDR implementation, such as land expropriation or exclusion from access to and benefits of these projects.

Overall, this obligation requires that climate adaptation and mitigation and similar actions are designed and implemented to avoid the threat to or breach of human rights. For instance, policies and laws related to the forest sector, climate finance, BECCS, and other land-use purposes must be designed and implemented in a manner that avoids land grabs and does not compromise food and water security. They must have due regard for vulnerable populations’ access to land and resources used for subsistence and other important cultural and economic purposes. States have an obligation to ensure that appropriate legislation and other measures are put in place to regulate the activities of non-state actors involved in the implementation of CDR. A key element of the duty to protect is the provision of effective, well-resourced, and technically equipped judicial and non-judicial mechanisms for receiving complaints and adjudicating disputes. These mechanisms should have the adequate jurisdiction to adjudicate cases, including those arising from alleged non-compliance with relevant standards and disputes involving extractive industries and local people. Furthermore, the relevant legislative act establishing such judicial mechanisms should allow citizens to pursue legal action and access effective remedies against states, companies, and other non-state actors whose actions or omissions undermine the climate system.

Similar to the obligation to protect human rights, according to the Africa Commission, is the obligation to promote the enjoyment of all human rights (SERAC v Nigeria, 2001). This obligation entails that the state should ensure ‘that individuals are able to exercise their rights, for example, by promoting tolerance, raising awareness, and even building infrastructures’ (SERAC v Nigeria, 2001). States in Africa have an obligation to promote by integrating climate change education with a focus on human rights in curricula at all levels of formal education. The obligation to promote includes raising awareness about climate change and human rights across sectors to ensure that the public has accessible, affordable, and understandable information regarding the benefits and risks of CDR. As an essential aspect of the obligation to promote, states in Africa must create national committees and commissions on climate geoengineering, especially CDR, where they do not already exist, to advance education and awareness about climate change, its adverse impacts, and associated loss and damages.

The obligation to fulfil, according to the African Commission, requires the state to mobilize ‘its machinery towards the actual realisation of the rights’ (SERAC v Nigeria, 2001, para 47). This includes taking steps to create the necessary conditions that enable people to safeguard the climate. Climate concerns should be integrated in all processes that empower right-holders to access, utilize, and develop natural resources. For example, this obligation encompasses establishing avenues for public and local communities to acquire shares in extractive industries (ACHPR, 2022).

In this context, states in Africa must mobilize funds and deploy revenue adequately in support of CDR initiatives, in so far as they enhance human rights. Locally and nationally generated funds for climate mitigation measures, such as carbon taxes, should serve a hypothecation purpose of boosting human rights in the climate domain (Nemavhidi & Jegede, 2023). The obligation to fulfil is a corollary of the obligation to cooperate among African states to address the challenges which climate change presents to state building in the continent. This includes sharing information about the science of climate change, African indigenous knowledge systems, and technologies and practices that improve understanding of climate geoengineering. States also have an obligation to seek international support, as climate change is a global crisis that requires collective action by the international community (Office of the United Nations High Commissioner for Human Rights, 2009). Financial and other forms of assistance are critical for the transfer of appropriate CDR technologies from developed states to Africa, building capacity, and investing in research.

5 Conclusion

There is increasing discussion on the potential of geoengineering as an intervention for addressing climate change, yet research has rarely focused on the involvement of the AU in this context. Therefore, with focus on the AU, this paper interrogated the trends in geoengineering, evaluated the adequacy of key regional legal framework, and explored how state human rights obligations may address the risks associated with the implementation of CDR options while enhancing human rights.

As demonstrated in the paper, the AU has rarely been involved in geoengineering projects. Yet its member states are endowed with resources which may be impacted by the agenda of geoengineering, particularly CDR options. As argued, there are opportunities and challenges in an Africa-wide implementation of afforestation, BECCS, and ocean fertilization that may both enhance and threaten key human rights, including the right to life, the right to have access to sufficient food and water, the right to have access to health care services, the right to property, the right to adequate housing, and the right to environment. These rights are guaranteed under the African Charter and supported by other AU legal and policy instruments relevant to geoengineering. The application of state obligations to 'respect', 'protect', 'promote', and 'fulfil' human rights can shape key AU legal and policy frameworks, contributing to minimizing anticipated risks while ensuring that the implementation of CDR geoengineering aligns with the promotion and protection of fundamental human rights across Africa.

References

ACHPR. (2000, May 11). *Malawi African Association and Others v Mauritania*. Communication No. 54/91, 61/91, 98/93, 164/97, 196/97, 210/98.

ACHPR. (2001). *Social and Economic Rights Action Center (SERAC) and Center for Economic and Social Rights (CESR) v Nigeria*. Communication 155/96.

ACHPR. (2013, November 5). *Front for the Liberation of the State of Cabinda v Republic of Angola*. Communication 328/06.

African Commission on Human and Peoples' Rights. (2008). Resolution on maternal mortality in Africa. In E. Durojaye & G. Mirugi-Mukundi (Eds.), *Compendium of documents and cases on the right to health under the African human rights system (2013)*.

African Commission on Human and Peoples' Rights. (2015). *General comment No. 3 on the African Charter on human and peoples' rights: The Right to Life (Article 4)*. Adopted during the 57th Ordinary Session of the African Commission on Human and Peoples' Rights held from 4 to 18 November 2015 in Banjul, The Gambia.

African Commission on Human and Peoples' Rights. (2022). *Background study on the operations of the extractive industries sector in Africa and its impacts on the realisation of human and peoples' rights under the African Charter on human and peoples' rights*. Adopted by the 69th Ordinary Session of the ACHPR. Available at: https://achpr.au.int/en/intersession-activity-reports/working-group-extractive-industries-environment-and-human-rights-3.

African Commission on Human and Peoples' Rights. (n.d.). *State parties to the African Charter*. https://achpr.au.int/en/states

African Development Bank. (2018). *Technologies for African agricultural transformation (TAAT)*. https://www.afdb.org/en/topics-and-sectors/initiatives-partnerships/technologies-african-agricultural-transformation-taat

African Heads of State and Government. (2023). *The African leaders Nairobi Declaration on climate change and call to Action*. Ministry of Finance, Government of Ghana.

African Union. (2003). *The African convention on the conservation of nature and natural resources* (Revised version) (Conservation Convention).

African Union. (2008). *Decision on the African common position on climate change*.

African Union. (2009). *African Union convention for the protection and assistance of internally displaced persons in Africa* (Kampala Convention).

African Union. (2014). *Agenda 2063, the Africa we want*.

African Union. (2022). *African Union climate change and resilient development strategy and action plan 2022–2032*.

Al-Zu'bi, M., Dejene, S. W., Hounkpè, J., Kupika, O. L., Lwasa, S., Mbege, M., Mwongera, C., Ouedraogo, N. S., & Toure, N. E. (2022). African perspectives on climate change research. *Nature Climate Change, 12*, 1078–1084.

Armani, M., Asefa, M., Zakari, S., & Agyekum, E. O. (2022). *Enhancing climate change adaptation and mitigation actions on land in Africa*. Future Ecosystems for Africa (FEFA). University of the Witwatersrand, Johannesburg.

AU-IBAR. (2019). *Africa blue economy strategy*. Nairobi, Kenya.

AUDA-NEPAD. (2014). *Framework and reform strategy for fisheries and aquaculture in Africa*.

Bellamy, R., Lezaun, J., & Palmer, J. (2017). Public perceptions of geoengineering research governance: An experimental approach. *Global Environmental Change, 45*, 81–90.

Burns, W. C. G. (2016). Human rights dimensions of bioenergy with carbon capture and storage: A framework for climate justice in the realm of climate geoengineering. In R. Abate (Ed.), *Climate justice: Case studies in global and regional governance challenges* (pp. 150–170). Environmental Law Institute.

Burns, W. C. G., & Nicholson, S. (2021). Bioenergy and carbon capture with storage (BECCS): The prospects and challenges of an emerging climate policy response. *Journal of Environmental Studies and Sciences, 7*, 527–534.

Centre for Minority Rights Development (Kenya) and Minority Rights Group (on behalf of Endorois Welfare Council). (2009, June–November). *Communication 276/03*. 27th Activity Report.
CESCR. (1991). *General Comment No. 4: The right to adequate housing (art. 11 (1) of the Covenant)*. 6th session.
CESCR. (1997). *General Comment No. 7: The right to adequate housing (art. 11 (1) of the Covenant): Forced evictions*. 16th session.
CESCR. (1999). *General Comment No. 12: The right to adequate food (art. 11)*. E/C.12/1999/5.
CESCR. (2000). *General Comment No. 14: The right to the highest attainable standard of health (art. 12)*. E/C.12/2000/4.
CESCR. (2002). *General Comment No. 15: The right to water (arts. 11 and 12 of the Covenant)*. E/C.12/2002/11.
Chalmin, A. (2021). Geoengineering activities on the African continent. *Geoengineering Monitor*. https://www.geoengineeringmonitor.org/2021/01/geoengineering-activities-on-the-african-continent/.
Ciais, P., Sabine, C., Bala, G., Bopp, L., Brovkin, V., Canadell, J., & Thornton, P. (2013). Carbon and other biogeochemical cycles. In *Climate change 2013: the physical science basis. Contribution of Working Group I to the Fifth Assessment Report of the Intergovernmental Panel on Climate Change* (pp. 465–570). Cambridge University Press.
Cotula, L, Oya, C., Codjoe, E. A., Eid, A., Kakraba-Ampeh, M., Keeley, J., Kidewa, A. L., Makwarimba, M., Seide, W. M., Nasha, W. O., Asare, R. O., & Rizzo, M. (2014). Testing claims about large land deals in Africa: Findings from a multi-country study. *Journal of Development Studies, 50*(7), 903–925.
Department of Minerals and Energy. (2007). *Biofuels industrial strategy of the Republic of South Africa*. https://www.energy.gov.za/files/esources/renewables/biofuels_indus_strat.pdf
Eick, M. V. (2010). A navigational system for uncharted waters: The London convention and London protocol's assessment framework on ocean iron fertilization. *Tulsa Law Review, 46*, 351–384.
FAO and AUDA-NEPAD. (2021). *Review of forest and landscape restoration in Africa 2021*. Accra.
Geoengineering Technology Briefing. (2021). *Marine cloud brightening or cloud reflectivity enhancement*. https://www.geoengineeringmonitor.org/wp-content/uploads/2021/04/marine-cloud-brightening.pdf
Gormley, W. P. (1987). The right to life in international law. *Denver Journal of International Law & Policy, 16*, 191–217.
Gush, M. B. (2006). Modelling streamflow reductions resulting from commercial afforestation in South Africa: From research to application. *Southern African Forestry Journal, 206*(1), 13–21.
Hale, B., & Dilling, L. (2011). Geoengineering, ocean fertilization, and the problem of permissible pollution. *Science, Technology, & Human Values, 36*(2), 190–212.
Hansungule, M., & Jegede, A. O. (2014). The impact of climate change on indigenous peoples' land tenure and use: The case for a regional policy in Africa. *International Journal on Minority and Group Rights, 21*(2), 241–274.
Hubbard, M. (2016). Barometer rising: The Cartagena protocol on biosafety as a model for holistic international regulation of ocean fertilization projects and other forms of geoengineering. *William & Mary Environmental Law and Policy Review, 40*, 591–636.
Human Rights Committee. (1982). *General Comment 6, Article 6* (Sixteenth session).
Human Rights Council. (various years). *Resolutions 7/23, 10/4, 18/22, 26/27, 29/15, 32/33, 35/20, and 38/4; A/HRC/10/61, A/HRC/32/23, A/HRC/35/13, A/HRC/38/21, A/HRC/41/26, A/64/255, A/67/299, A/HRC/36/46, A/HRC/41/39, A/HRC/16/43, A/66/285, A/HRC/31/52, A/HRC/25/53.*
ICJ. (1996). *Advisory opinion on the legality of the threat or use of nuclear weapons*. I.C.J. Rep. 226.
ICJ. (2010). *Case concerning pulp mills on the River Uruguay (Argentina v Uruguay)*. Judgment, I.C.J. Rep. 14.
IHRDA and Others v. DRC. (2016). *Communication 393/10* (Kilwa decision).
IPCC. (2023). Summary for policymakers. In *Climate Change 2023: Synthesis Report. Contribution of Working Groups I, II and III to the Sixth Assessment Report of the Intergovernmental Panel on*

Climate Change [Core Writing Team, H. Lee and J. Romero (eds.)]. IPCC, Geneva, Switzerland, 1–34.
IUCN. (2004). *Afforestation and reforestation for climate change mitigation: Potentials for Pan-European action.* https://foresteurope.org/wp-content/uploads/2016/08/Afforestation.pdf
Jegede, A. O. (2016). Rights away from home: Climate-induced displacement of indigenous peoples and the extraterritorial application of the Kampala Convention. *African Human Rights Law Journal, 16*(1), 58–82.
Jegede, A. O. (2021). The role of human rights in implementing CDR geoengineering options in South Africa. In W. Burns, D. Dana, & S. J. Nicholson (Eds.), *Climate geoengineering: Science, law and governance* (pp. 145–163). Springer International Publishing.
Jonker, W., Alan, B., Musango, C., Kaviti, J., & de Kock, I. (2017). Implications of biofuel production in the Western Cape province, South Africa: A system dynamics modelling approach. *Journal of Energy in Southern Africa, 28*(1), 1–12.
Kartha, S., & Dooley, K. (2016). *The risks of relying on tomorrow's 'negative emissions' to guide today's mitigation action* (SEI Working Paper No. 2016-08).
Keane, K. (2020). Geo-engineering the climate: A preliminary examination of international governance challenges and opportunities. *Trinity College Law Review, 23*, 56–89.
Koornneef, J., van Breevoort, P., Hamelinck, C., Hendriks, C., Hoogwijk, M., Koop, K., Koper, M., Dixon, T., & Camps, A. (2012). Global potential for biomass and carbon dioxide capture, transport and storage up to 2050. *International Journal of Greenhouse Gas Control, 11*, 117–132.
Lampitt, R. S., Achterberg, E. P., Anderson, T. R., Hughes, J. A., Iglesias-Rodriguez, M. D., Kelly-Gerreyn, B. A., Lucas, M., Popova, E. E., Sanders, R., Shepherd, J. G., Smythe-Wright, D., & Yool, A. (2008). Ocean fertilization: A potential means of geoengineering? *Philosophical Transactions of the Royal Society A: Mathematical, Physical and Engineering Sciences, 366*(1882), 3919–3945.
Malkamaki, A., D'Amato, D. J., Hogarth, N., Kanninen, M., Pirard, R., Toppinen, A., & Zhou, W. (2018). A systematic review of the socio-economic impacts of large-scale tree plantations, worldwide. *Global Environmental Change, 53*, 90–103.
Marshall, J. B. (2017). Geoengineering: A promising weapon or an unregulated disaster In the fight against climate change? *Florida State University Journal of Land Use and Environmental Law, 33*(1), 183–214.
Morrow, R. D., Thompson, S. M., Anderson, A., Batres, M., Buck, J. H., Dooley, K., Geden, O., Ghosh, A., Low, S., Njamnshi, A., Noë, J., Femi, O., Ta, O., Talati, S., & Wilcox, J. (2020). Principles for thinking about carbon dioxide removal in just climate policy. *One Earth, 3*, 150–153.
Nemavhidi, M., & Jegede, A. O. (2023). Carbon tax as a climate intervention in South Africa: A potential aid or hindrance to human rights? *Environmental Law Review, 25*(1), 11–32.
Nowak, D. J., Hoehn, R., & Crane, D. E. (2007). Oxygen production by urban trees in the United States. *Arboriculture & Urban Forestry, 33*(3), 220–226.
OAU. (1981). *African Charter on human and peoples' rights.* CAB/LEG/67/3 rev. 5, 21 ILM 58.
Office of the United Nations High Commissioner for Human Rights. (2009). *Report of the Office of the United Nations High Commissioner for Human Rights on the Relationship Between Climate Change and Human Rights.* A/HRC/10/61.
Paris Agreement. (2015, December 12). *Paris Agreement to the United Nations framework convention on climate change* (T.I.A.S. No. 16-1104).
Shue, H. (1980). *Basic rights: Subsistence, affluence, and U.S. Foreign Policy* (2nd ed.). Princeton University Press.
Singh, S. K., Kotakonda, A., Kapardar, R. K., Ramaiah, N., & Amann, R. (2015). Response of bacterioplankton to iron fertilization of the Southern Ocean, Antarctica. *Frontiers in Microbiology, 6*, 1–13.
Smith, P. (2016). Soil carbon sequestration and biochar as negative emission technologies. *Global Change Biology, 22*(3), 1315–1324.

Smith, P., Davis, S. J., Creutzig, F., Fuss, S., Minx, J., Gabrielle, B., Kato, E., Jackson, R. B., Cowie, A., Kriegler, E., van Vuuren, D. P., Rogelj, J., Ciais P., Milne, J., Canadell, J. G., McCollum, D., Peters, G., Andrew, R., Krey, V., …, Yongsung, C. (2016). Biophysical and economic limits to negative CO_2 emissions. *Nature Climate Change, 6*, 42–50.

Stocker, T., Qin, D., Plattner, G. K., Tignor, M., Allen, S. K., Boschung, J., Nauels, A., Xia, Y., Bex, V., & Midgley, P. M. (2013). *The physical science basis.* Contribution of Working Group I to the 5th Assessment Report of the Intergovernmental Panel on Climate Change (IPCC Summary for Policy Makers). Cambridge University Press.

Svoboda, T., Buck, H. J., & Suarez, P. (2019). Climate engineering and human rights. *Environmental Politics, 28*(3), 397–416.

Thangaraj, P., Okoye, S., Gordon, B., Zilberman, D., & Hochman, G. (2018). *Fact sheet: Bioenergy with carbon capture and storage.* Department of Agricultural and Resource Economics.

The Royal Society. (2009). *Geoengineering the climate: Science.* The Royal Society.

The Royal Society & Royal Academy of Engineering. (2018). *Greenhouse gas removal.* The Royal Society.

Trail Smelter Case (United States v. Canada). (1941). 3 R.I.A.A. 1905.

Turner-Skoff, J. B., & Cavender, N. (2019). The benefits of trees for livable and sustainable communities. *Plants, People, Planet, 1*, 323–335.

Turpie, J. K., O'Connor, T., Mills, A., & Robertson, H. (2007). The ecological and economic consequences of changing land use in the Southern Drakensberg Grasslands, South Africa. *South African Journal of Economic and Management Sciences, 10*(4), 423–441.

United Nations. (2023). *Global water security 2023 assessment.* https://reliefweb.int/report/world/global-water-security-2023-assessment

Reflections on Geoengineering and Biodiversity

Perses Bilimoria

Abstract Biodiversity, the variety of life forms on Earth, is crucial for maintaining ecosystem balance and is currently threatened by human-induced climate change. Geoengineering, involving large-scale interventions like Carbon Dioxide Removal (CDR) and Solar Radiation Management (SRM), aims to restore this equilibrium. As global warming risks exceeding 1.5 °C, rapid biodiversity loss and ecosystem disruption become imminent. Geoengineering solutions, though promising, carry significant risks, including potential ozone depletion and unforeseen ecological impacts. Effective governance, involving both state and non-state actors, and robust policy frameworks are essential to ensure these interventions support biodiversity and global climate goals without causing further harm.

Keywords Geoengineering · Biodiversity · CDR · SRM · Climate change

1 Importance and Relevance of Biodiversity for Geoengineering

Biodiversity encompasses the incredible variety of life forms on Earth including, plants, bacteria, animals, fungi, and humans. It is estimated that our planet hosts around 8.7 million species (Mora et al., 2011), yet only about 1.2 million have been identified and described so far, with insects comprising the majority of known species (Mora et al., 2011). However, every single species plays a vital role in maintaining the delicate balance of ecosystems.

Geoengineering, also known as climate intervention, refers to methods deployed at scale to modify Earth's climate system, including interventions in the atmosphere, oceans, and terrestrial systems (National Research Council, 2015). These interventions are being considered due to the widespread impact of human-induced warming and the challenge of rapidly reducing greenhouse gas emissions. Climate intervention

P. Bilimoria (✉)
Judge Business School, Cambridge Institute for Sustainability Leadership, Cambridge University, Cambridge, UK
e-mail: pbilimoria@gmail.com

K.-U. Schrogl et al. (eds.), *The Politics of Geoengineering*, Springer Climate,
https://doi.org/10.1007/978-3-032-10095-5_7

or geoengineering incorporates the application of scientific technologies to address the challenge of anthropogenic climate change.

The urgency of the situation becomes apparent when we consider that global warming continues to approach 1.5 °C above pre-industrial levels (IPCC, 2023). Should current warming trends continue in a non-linear fashion, we may witness rapid destruction and extinction of significant biodiversity and ecosystem disruption that supports both planetary boundaries (Rockström et al., 2009) and humanity. Such a scenario would signify a transition of our Earth system from the relative stability of the Holocene, during which human civilization developed, to the uncertain conditions of the Anthropocene (Crutzen, 2002).

Responding to this urgency has prompted research into climate intervention solutions such as Carbon Dioxide Removal (CDR) and Solar Radiation Management (SRM) (National Research Council, 2015). Reflecting on recent technological advances, humanity has demonstrated capacity to develop and deploy new technologies rapidly when facing existential threats. Similarly, we must evaluate climate intervention technologies that, although requiring further validation, may provide pathways toward sustainable climate management within limited timeframes.

The concept of weather and climate modification research dates back to the early 1950s, when it became a focus of atmospheric sciences research (Fleming, 2010). While the world has focused on carbon mitigation and sequestration as primary solutions to climate change, it is important to understand that atmospheric water vapor, while the most abundant greenhouse gas, acts primarily as a feedback mechanism rather than a forcing agent (NASA, 2024). Water vapor contributes approximately 50% of the natural greenhouse effect (Dessler et al., 2013), but its concentration responds to temperature changes driven by long-lived greenhouse gases like CO_2, methane, and nitrous oxide (Schmidt et al., 2010). The concentration of water vapor in the atmosphere is regulated by air temperature through the Clausius-Clapeyron relation, creating a positive feedback loop that amplifies warming (Sherwood et al., 2010).

Therefore, while water vapor plays a crucial amplifying role in climate change, the primary drivers of anthropogenic warming remain the long-lived greenhouse gases directly emitted by human activities (IPCC, 2023). The intricate interplay between biodiversity, geoengineering, and climate feedbacks will determine our trajectory into an uncertain future.

2 Defining Key Problems and Conflicts

Undertaking climate intervention entails inherent risks, encompassing both natural and societal dimensions. For example, important questions exist regarding how stratospheric aerosols could affect ozone concentrations (Robock et al., 2009). Furthermore, questions emerge about whether research into climate intervention might reduce incentives for emissions mitigation (McLaren, 2016).

The environmental and ecological risks differ considerably for both CDR and SRM approaches. In both cases, the risks would be twofold: firstly, the risk of unintended side effects locally and globally, and secondly, the risk that intended outcomes would not materialize (Trisos et al., 2018). It is vital to recognize that rapid implementation and subsequent termination of both SRM and CDR could hold potentially dangerous consequences for biodiversity should the initiatives prove unsuccessful (Trisos et al., 2018).

For instance, within SRM approaches, concerns exist regarding effects on ozone concentrations, as well as risks associated with termination problems given the difficulty in predicting climatic responses from albedo modifications (McClellan et al., 2012; Robock et al., 2009). Similarly, CDR deployment presents potential side effects, such as possible biodiversity loss or reduced ecosystem services, which could arise from large-scale ecosystem manipulation for carbon sequestration (Smith et al., 2016). Additionally, risks exist regarding the permanence of carbon storage and potential for rapid re-release of sequestered carbon (Fuss et al., 2018).

However, it is crucial to acknowledge that the environmental and geophysical risks linked to climate intervention are not yet fully understood. Consequently, decisions regarding research and potential deployment depend on assessments of scalability, reversibility, and governance frameworks (Parker & Irvine, 2018). Can these approaches be tested safely on small scales? Should they prove inadequate or harmful, can they be readily reversed or terminated?

Given the complexity of these interventions, robust governance frameworks are essential. These initiatives must be evaluated based on scientific evidence and their compatibility with biodiversity conservation and sustainable development goals (Chhetri et al., 2018).

3 Actors and Institutions Involved

In response to concerns about climate change impacts and insufficient progress on emissions reductions, various actors are engaging with geoengineering research and governance discussions. These actors include researchers and research institutions, governmental agencies, non-governmental organizations (NGOs), professional societies, philanthropic funders, and academic publishers engaged in advancing both SRM and CDR research (Reynolds & Parson, 2020).

Given expanding engagement in this field, the importance of research governance becomes crucial. Key principles include scientific rigor, transparency, public engagement, and international cooperation, especially as research advances toward larger-scale studies (Parker & Irvine, 2018).

Among philanthropic research funders, organizations such as the Bill and Melinda Gates Foundation, Grantham Foundation, and others have supported climate intervention research (Reynolds & Parson, 2020). However, funding patterns and priorities continue to evolve as the field develops.

On institutional levels, the United Nations Environment Programme (UNEP) and other international bodies are beginning to engage with geoengineering governance questions, though formal frameworks remain limited (UNEP, 2023).

It is important to acknowledge that governance challenges exist regarding legitimacy, accountability, and representation in geoengineering research and potential deployment decisions (Biermann & Möller, 2019). These challenges highlight the need for inclusive governance approaches involving multiple stakeholders and robust public engagement processes.

In light of these challenges, enhanced cooperation between state and non-state actors, combined with transparent governance frameworks, could help address concerns while fostering responsible research and development approaches (Parker & Irvine, 2018).

4 Policy and Regulatory Approaches with Possible Solutions

The development of regulatory frameworks for climate intervention must consider relevance to existing international agreements, particularly the Convention on Biological Diversity (CBD).

The CBD has issued guidance urging parties and governments to ensure that climate-related geoengineering activities that may affect biodiversity are not undertaken "in the absence of science-based, global, transparent and effective control and regulatory mechanisms" (CBD, 2016). Limited exceptions exist for small-scale research activities conducted under appropriate governance frameworks.

Additionally, while the precautionary principle provides relevant guidance, its precise application to emerging technologies like geoengineering requires careful consideration within existing legal frameworks (Barrett, 2008). The UNFCCC continues to address these complex governance questions as understanding of the technologies develops.

Developing regulatory frameworks for climate intervention must consider existing treaty regimes and institutions, including:

- The Convention on Biological Diversity (CBD)—Entered into force December 29, 1993.
- The United Nations Convention on the Law of the Sea (UNCLOS)—Entered into force November 16, 1994
- The London Convention and Protocol on marine pollution prevention—LC entered into force August 30, 1975; LP entered into force March 24, 2006.
- The Kyoto Protocol to the UNFCCC—Entered into force February 16, 2005.
- The Montreal Protocol on Ozone Layer Protection—Entered into force January 1, 1989.
- The Environmental Modification Convention (ENMOD)—Entered into force October 5, 1978.

- Space law frameworks relevant to potential space-based SRM approaches.
- The OSPAR Convention for North-East Atlantic marine environment protection—Entered into force March 25, 1998.
- The Convention on Long-Range Transboundary Air Pollution—Opened for signature November 13, 1979
- The Antarctic Treaty—Entered into force June 23, 1961.

International institutions including the UN General Assembly, UNEP, World Meteorological Organization, and Intergovernmental Oceanographic Commission all have relevance to climate intervention governance questions.

Given governance gaps in current frameworks, some scholars have proposed establishing dedicated international oversight mechanisms, potentially modeled on existing institutions like the International Seabed Authority, to coordinate global research and any future deployment decisions (Biermann & Möller, 2019). Simultaneously, frameworks for research governance at local and regional levels could help ensure responsible development of these technologies.

5 Exploring Policy and Diplomatic Mechanisms to Reach Solutions

Contemporary climate change represents an unprecedented challenge requiring innovative responses. Throughout history, humanity has addressed major challenges through technological innovation combined with international cooperation.

To address accelerating climate change, technological approaches including climate intervention may play important roles alongside aggressive emissions reductions. Such technologies could provide critical options for managing climate risks while society transitions to sustainable energy systems.

The importance of international cooperation has been demonstrated through agreements like the Paris Agreement, where nations committed to limiting warming to well below 2 °C while pursuing efforts to limit warming to 1.5 °C (UNFCCC, 2015). However, current national commitments remain insufficient to achieve these temperature goals (UNEP, 2023).

Given the urgency of climate action and the scale of required changes, research into climate intervention approaches becomes increasingly relevant for comprehensive climate policy portfolios. However, such research must be conducted responsibly with appropriate governance frameworks.

Our collective efforts should focus on developing policy approaches and regulations based on robust science and precautionary principles. These frameworks should ensure that any climate intervention research or deployment protects existing biodiversity while supporting broader sustainable development objectives.

References

Barrett, S. (2008). The incredible economics of geoengineering. *Environmental and Resource Economics, 39*(1), 45–54.

Biermann, F., & Möller, I. (2019). Rich man's solution? Climate engineering discourses and the marginalization of the Global South. *International Environmental Agreements, 19*(2), 151–167.

CBD (Convention on Biological Diversity). (2016). *Decision XIII/21: Climate-related geoengineering.* Conference of the Parties to the Convention on Biological Diversity, Thirteenth meeting, Cancun, Mexico.

Chhetri, N., Chong, D., Conca, K., Falk, R., Gillespie, A., Gupta, A., Jinnah, S., Kashwan, P., Lahsen, M., Light, A., & Nicholson, S. (2018). *Governing solar radiation management.* Forum for Climate Engineering Assessment.

Crutzen, P. J. (2002). Geology of mankind. *Nature, 415*(6867), 23.

Dessler, A. E., Schoeberl, M. R., Wang, T., Davis, S. M., & Rosenlof, K. H. (2013). Stratospheric water vapor feedback. *Proceedings of the National Academy of Sciences, 110*(45), 18087–18091.

Fleming, J. R. (2010). *Fixing the sky: The checkered history of weather and climate control.* Columbia University Press.

Fuss, S., Lamb, W. F., Callaghan, M. W., Hilaire, J., Creutzig, F., Amann, T., Beringer, T., de Oliveira Garcia, W., Hartmann, J., Khanna, T., & Minx, J. C. (2018). Negative emissions—Part 2: Costs, potentials and side effects. *Environmental Research Letters, 13*(6), 063002.

IPCC. (2023). *Climate Change 2023: Synthesis Report. Contribution of Working Groups I, II and III to the Sixth Assessment Report of the Intergovernmental Panel on Climate Change* [Core Writing Team, H. Lee and J. Romero (eds.)]. IPCC, Geneva, Switzerland.

McClellan, J., Keith, D. W., & Apt, J. (2012). Cost analysis of stratospheric albedo modification delivery systems. *Environmental Research Letters, 7*(3), Article 034019.

McLaren, D. (2016). Mitigation deterrence and the "moral hazard" of solar radiation management. *Earth's Future, 4*(12), 596–602.

Mora, C., Tittensor, D. P., Adl, S., Simpson, A. G., & Worm, B. (2011). How many species are there on Earth and in the ocean? *PLoS Biology, 9*(8), Article e1001127.

NASA. (2024). Steamy relationships: How atmospheric water vapor amplifies Earth's greenhouse effect. *NASA Science.* https://science.nasa.gov/earth/climate-change/steamy-relationships/

National Research Council. (2015). *Climate intervention: Carbon dioxide removal and reliable sequestration.* The National Academies Press.

Parker, A., & Irvine, P. J. (2018). The risk of termination shock from solar geoengineering. *Earth's Future, 6*(3), 456–467.

Reynolds, J. L., & Parson, E. A. (2020). Nonstate governance of solar geoengineering research. *Climate Policy, 20*(3), 314–334.

Robock, A., Marquardt, A., Kravitz, B., & Stenchikov, G. (2009). Benefits, risks, and costs of stratospheric geoengineering. *Geophysical Research Letters, 36*(19).

Rockström, J., Steffen, W., Noone, K., Persson, Å., Chapin III, F. S., Lambin, E. F., Lenton, T. M., Scheffer, M., Folke, C., Schellnhuber, H. J., & Foley, J. A. (2009). A safe operating space for humanity. *Nature, 461*(7263), 472-475.

Schmidt, G. A., Ruedy, R. A., Miller, R. L., & Lacis, A. A. (2010). Attribution of the present-day total greenhouse effect. *Journal of Geophysical Research: Atmospheres, 115*(D20).

Sherwood, S. C., Roca, R., Weckwerth, T. M., & Andronova, N. G. (2010). Tropospheric water vapor, convection, and climate. *Reviews of Geophysics, 48*(2).

Smith, P., Davis, S. J., Creutzig, F., Fuss, S., Minx, J., Gabrielle, B., Kato, E., Jackson, R. B., Cowie, A., Kriegler, E., & Yongsung, C. (2016). Biophysical and economic limits to negative CO_2 emissions. *Nature Climate Change, 6*(1), 42-50.

Trisos, C. H., Amatulli, G., Gurevitch, J., Robock, A., Xia, L., & Zambri, B. (2018). Potentially dangerous consequences for biodiversity of solar geoengineering implementation and termination. *Nature Ecology & Evolution, 2*(3), 475–482.

UNFCCC. (2015). *Paris Agreement*. United Nations Framework Convention on Climate Change.
UNEP. (2023). *Emissions Gap Report 2023: Broken Record—Temperatures hit new highs, yet world fails to cut emissions (again)*. United Nations Environment Programme.

Geoengineering and International Law

Alla Pozdnakova

Abstract This chapter examines the role of international law in the emerging governance landscape of GE. Key issues of international law on GE remain debatable and unresolved. It is argued that internationally harmonized standards for GE are indispensable for the effective governance of these activities. A prospective GE regime needs to be comprehensive, multilateral, and backed-up by appropriate institutional framework(s). It also needs to incorporate non-state perspectives and promote coherence between applicable international law rules. The legal research community, in collaboration with other sciences, faces a formidable task of promoting scientifically informed decision-making and working out feasible and effective legal solutions to fill out gaps for GE.

Keywords Geoengineering · International law · Marine dumping · Precautionary principle · BBNJ

1 Introduction

Geoengineering (GE) is a generic term embracing activities aimed at deliberate large-scale manipulation of the Earth's environment to alter the climate system and alleviate the impacts of climate change (Lerner, 2021). Geoengineering activities are commonly divided into two broad categories: solar geoengineering, or Solar Radiation Modification (SRM), and Carbon Dioxide Removal (CDR).[1] The Intergovernmental Panel on Climate Change (IPCC) 6th Assessment Report distinguishes clearly between CDR and SRM, due to their very different geophysical characteristics (IPCC, 2022, p. 168).

Despite important differences between various GE techniques and approaches, they share some fundamental features. Firstly, GE does not deal with the problem at its source, that is, it does not seek to accomplish reductions of greenhouse gas (GHG) emissions. GE techniques only tackle consequences of the global warming

A. Pozdnakova (✉)
Faculty of Law, Scandinavian Institute of Maritime Law, University of Oslo, Oslo, Norway
e-mail: alla.pozdnakova@jus.uio.no

K.-U. Schrogl et al. (eds.), *The Politics of Geoengineering*, Springer Climate,
https://doi.org/10.1007/978-3-032-10095-5_8

by temporarily cooling the Earth on a global or regional basis. Nevertheless, GE is a potential solution for this issue of global concern. At some point in future, application of GE to 'turn the thermostat down' may become inevitable as a measure of last resort.[2]

Secondly, GE is technology-intensive and increasingly commercialized (Chalmin, 2023). Already in foreseeable future, it is likely to evolve from an experimental, scientist-driven, and small-scale into commercial, market-driven activity. Thirdly, an inherently transboundary nature of GE—and the impact it may have on global climate and environment—makes it a matter of high concern for the international community of States and for the humankind. The pertinence of international legal solutions to govern GE has been highlighted in scientific reports, academic literature and statements by international organizations (GESAMP, 2019; IPCC, 2022; Proelss, 2012; UNEP, 2023).

GE is not presently regulated by a comprehensive, uniform, and coherent international legal regime but by a fragmented patchwork of international legal principles and rules (GESAMP, 2019; Proelss, 2012; UK-DNVA, 2023). International law development on GE should be seen in a broader international law landscape, as GE regulation evolves in a close interplay with other areas of international law (climate change and environmental protection, global commons, human rights, etc.). International law developments on GE and ecocide and GE and human rights are meticulously examined elsewhere in this book.[3]

This chapter critically examines the role of international law in the emerging governance landscape of GE. It begins with briefly outlining the existing international law and governance frameworks for GE with focus on sectoral approaches by the parties to the Convention on Biological Diversity (CBD)[4] and the International Maritime Organization (IMO) (Sect. 2). Then it explores the effectiveness of international legal approaches to governance of GE. While recognizing that such governance must be based on interdisciplinary approaches and mechanisms,[5] this author argues that international law plays an indispensable role in the effective and equitable governance and regulation of GE (Sect. 3). Section 4 dwells on those international legal aspects of the future GE governance which, in this author's view, require further attention from the research community. Conclusions are included in Sect. 5.

2 The State of International Law Development on GE: Selected Examples

No general prohibition of GE is found in international law, but legal scholars point out that deployment of GE techniques may violate fundamental principles and obligations of international environmental law such as precautionary principle, no-harm duty, and the duty to notify and consult (Brent, 2021; Proelss, 2012, pp. 205–211). Several states, international organizations, scientific bodies, and local communities also have expressed skepticism toward GE (Gupta, 2016). However, as the international community predictably fails to reduce GHG emissions, development and application of GE remain on the table and even continues to grow, despite concerns with feasibility and safety of various GE techniques (Njagi, 2024). Experimental and small-scale GE projects have already been conducted, even in the absence of a universally recognized international law and governance framework (Lee, 2011; UARCTIC, 2024). Thus, a real question is not whether GE is prohibited or precluded by international law. In future, we should be prepared to answer whether the application of GE techniques is required by international law as a measure of last resort (Corbett, 2022).

The IPCC defines geoengineering as a 'broad set of methods and technologies that aim to deliberately alter the climate system in order to alleviate the impacts of climate change' (IPCC, 2011, p. 11). A central point of agreement across scientific communities is that effects of GE techniques are scientifically and ecologically uncertain and require an approach based on precaution (Gupta, 2016). The two major doubts lie, firstly, in our ability to correctly assess adverse impacts of GE and, secondly, in GE's potential for the exacerbation of related environmental issues (Royal Society, 2024, p. 18).

Despite its unique and novel characteristics, GE has already been regulated by international measures. The two important examples of sectoral instruments are provided by the decisions of the Conference of the Parties (COP) to CBD and by member States of the IMO. Despite their different mandates and competences, both bodies have expressed largely similar concerns with GE and acknowledged each other's work (Gupta, 2016). Notably, both CBD and IMO instruments call for a precautionary approach to GE (Steenkamp, 2025).

CBD's preamble requires that 'lack of full scientific certainty should not be used as a reason for postponing measures to avoid or minimize' 'a threat of significant reduction or loss of biological diversity'. This Convention also sets out provisions on the environmental impact assessment and minimizing adverse impacts of proposed projects that are likely to have *significant adverse effects* on biological diversity (Article 14).

The 2008-moratorium on ocean fertilization adopted by the COP to CBD reiterates the relevance of the precautionary approach and calls for States to refrain from ocean fertilization 'until there is an adequate scientific basis on which to justify such activities'.[6] The moratorium was expanded in 2010 to include climate change-related *large-scale* geoengineering activities that may affect biodiversity and adds the need for 'appropriate consideration of the associated risks for the environment and biodiversity and associated social, economic and cultural impacts'.[7] Albeit voluntary, non-binding international commitments, the 2008- and 2010-decisions are still important as expressions of the common will of CBD States Parties. The moratorium may be implemented in a binding form by national measures (Morgera, 2011, p. 98).

The IMO has addressed GE in the context of marine dumping and disposal of wastes at sea regulated in the London Convention on Dumping (the London Convention, or LC)[8] and the 1996 Protocol (LP) (Scott, 2023).[9] Unlike CBD, which prescribes for States Parties to take measures to protect biodiversity (of which a GE-moratorium is but one example), the LC/LP and their related instruments lay down an outright prohibition of marine dumping. An exception is allowed for dumping of specifically enlisted wastes, but it may only be conducted subject to permits issued by the relevant national authorities.[10] The LC/LP expressly envisages a precautionary approach to all marine dumping activities included in their scope.

The LC/LP does not expressly refer to GE but recognizes that some of GE techniques share common features with the expressly included dumping activities. It is pointed out in literature that LC has from the outset had a broad scope, enabling it to encompass more than dumping at sea (Scott, 2023). As the account below shows, the London dumping framework has continuously evolved to address novel activities (albeit not necessarily in a binding form), proving that a treaty regime is capable of being forward-looking and dynamic.

In 2006, LC/LP (Annex I) was amended to regulate offshore CO^2 sequestration as a climate change mitigation measure (Scott, 2023, p. 241). In 2008, the LC/LP framework was supplemented by a resolution which includes ocean fertilization activities into the LC/LP framework.[11] Rather than by highlighting *large-scale* activities as a matter of concern, the 2008 LC/LP-resolution differentiates between *commercial* ocean fertilization activities and *scientific* research involving ocean fertilization. It is pointed out that current knowledge on the effectiveness and potential environmental impacts of ocean fertilization is insufficient to justify activities other than legitimate scientific research on ocean fertilization. In 2010, IMO member States adopted an Assessment Framework for Scientific Research Involving Ocean Fertilization to determine, 'with utmost caution', whether a proposed ocean fertilization activity constitutes legitimate scientific research that is not contrary to the aims of the LC/LP.[12]

In 2013, the LP was amended to include marine geoengineering (Bohman & Ringbom, 2023; Scott, 2023).[13] The LP defines 'marine geoengineering' as a 'deliberate intervention in the marine environment to manipulate natural processes, including to counteract anthropogenic climate change and/or its impacts, and that has the *potential* to result in deleterious effects, *especially* [but not exclusively—A.P.] where those effects may be *widespread, long-lasting or severe*' (author's italics).[14] The 2013-amendment to LP stands out among all other international instruments on GE: once in force, it will become 'the first (and potentially only) global legally binding regime to explicitly regulate marine geoengineering for climate purposes' (Scott, 2023, p. 241).

In 2022, a Statement on Marine Geoengineering was adopted by States parties to LC/LP to encompass SRM techniques: marine cloud brightening and microbubbles/reflective particles/material.[15] Regulation of these techniques is currently not included in the LC/LP, but IMO is conducting a legal and technical analysis to evaluate options for future action.

The relevance of the precautionary approach is also discernible in the international instruments on the protection of the atmosphere, e.g., the Montreal Protocol[16] to the Vienna Convention on the Protection of the Ozone Layer.[17] With regard to intentional large-scale modification of the atmosphere, the International Law Commission (ILC)

Draft Guidelines on Protection of the Atmosphere provide that such activities 'should only be conducted with prudence and caution, and subject to any applicable rules of international law, including those relating to environmental impact assessment'.[18] The ILC Draft Guidelines state that they do not deal with and are without prejudice to the polluter-pays principle, the precautionary principle and the common but differentiated responsibilities principle.[19] Albeit indirectly, the ILC acknowledges applicability of environmental principles to SRM projects. Regrettably, the resistance to fully recognize and substantiate these principles for the protection of the atmosphere illustrates that the international legal framework governing GE is debatable and far from being fully agreed to by States.

3 An Appraisal of the Contribution of International Law to the Governance of GE

The CBD and IMO instruments provide an illustration of the international law's contribution to the shaping of GE governance by binding and 'soft' multilateral obligations for States to take precautionary measures and to apply specific evaluation criteria when authorizing GE projects. Thus, the international legal order constitutes an important component in the effective governance of GE because it shapes and develops legal norms for States (and, through national implementation, non-State actors). However, as argued in this section, international law has a much broader arsenal of solutions and tools to address GE, which has so far remained largely unused.

Generally, international law is slow to respond to technology-driven and rapidly evolving new sectors, GE not being an exception. It has been early recognized that a multilateral, cooperation-based approach is indispensable to regulate GE, an activity whose environmental impact is uncertain and which may potentially be used in a hostile manner (Mayer, 2018, pp. 147–148).[20] A universal, science-based approach to studies of large-scale weather modification was proposed by UNGA already in 1961,[21] followed by the statement by UNEP on the duty to cooperate in a spirit of good neighborliness with regard to weather modification projects (Mayer, 2018, p. 148). However, the international community has not yet reached a universal consensus on what an international governance framework for GE should be like.

The early initiatives of the 1960s–early 1980s have not been consistently followed up in the UN-fora, including the UN Framework Convention on Climate Change (UNFCCC), due to significant doubts dominating the discourse on the feasibility of GE techniques (Mayer, 2018, p. 157). The UN Committee on Peaceful Uses of Outer Space (COPUOS) has also largely focused on the application of space technologies and applications for emission monitoring and control, rather than on legal solutions for SRM (COPUOS, 2024).[22] Despite the current realities of GE, many States resist even a proposal to initiate a multilateral work on GE. For example, at the UN Environmental Assembly (UNEA) in 2019, and repeatedly in 2024, Switzerland presented a proposal for a UNEA-resolution on SRM, justifying it by concerns with the lack of multilateral control and oversight over SRM (Watts, 2024). The resolution has been rejected both times by participating States (Njagi, 2024).

Thus, the main progress toward multilateral GE governance have so far been accomplished by the *sectoral* measures of CBD and the IMO described earlier in Sect. 2. These examples serve to illustrate that GE governance evolves with the growing awareness about the potential risks and uncertainties of particular GE projects. An important lesson is also that the promotion of new legal instruments and provisions is facilitated by the existence of institutional frameworks which enable cooperation and dialogue among participating States (Kravik, 2021). It is remarkable that, within the respective scopes of the underlying conventions, States Parties to CBD and the IMO have managed to adopt a prohibition or 'moratorium' on GE, subject to certain criteria and conditions. Their central contribution is, in this author's view, the confirmation that the precautionary principle (and approach) is central for the regulation of GE. Notably, the 2013-amendment to LC/LP on marine GE also details the criteria for the application of the precautionary approach as it introduces provisions on the assessment of potential effects, risk management, monitoring, and permit conditions.

However, despite their remarkable achievements on GE, sectoral organizations—even global ones—may not on their own meaningfully engage with the GE governance. GE, inherently liable to create lasting effects for both the atmosphere, biodiversity, oceans, and climate, should be governed by an international legal regime that is capable of tackling the interactions and interdependencies of these spaces (Brunnée, 2019). This calls for an UN-level, universal development of principles and rules governing GE.

Furthermore, while law and legal instruments and norms are essential for an effective GE governance, their contribution may only be realized if backed-up by the interdisciplinary insights, approaches, stakeholders, and institutions. It is crucial to develop and apply legal norms for environmental protection on the basis of scientific knowledge.[23] A critical and persistent issue in the international law is how to create and operationalize such a co-opted regime which is capable of effectively incorporating these complex dimensions (Low et al., 2013, p. 188).

The UN Agreement under the UN Convention on the Law of the Sea on the conservation and sustainable use of marine biological diversity of the areas beyond national jurisdiction (BBNJ) is a fresh example of an international, binding framework agreement which attempts to cut across the established sectoral and regional divisions and to transcend the traditional boundaries of statehoods by including a broader range of stakeholders (including the indigenous communities).[24] The substantive reach of the BBNJ is broad and applies to all activities with regard to the areas beyond national jurisdiction (ABNJ). The precise geographic scope of the BBNJ is determined in accordance to UNCLOS provisions which contain 'negative' definitions of the high seas (Article 87) and the Area (Article 1(1)). Arguably, activities conducted in the atmosphere are also covered by the BBNJ to the extent it is necessary for protection of marine biodiversity in ABNJ. The Agreement recognizes the ocean-climate nexus (Voigt, 2023).

In this author's view, the BBNJ is a crucial multilateral legal instrument in the emerging GE governance. The BBNJ re-affirms the relevance of environmental law principles such as precautionary principle (or approach), ecosystem approach, and the use of the best available science and scientific information (Article 7). Importantly, it seeks to build bridges between existing environmental regimes and norms, by creating an institutional framework for cooperation and coordination by global and regional organizations. BBNJ also envisages creation of new, self-standing, institutions to be established under this Agreement, including Scientific and Technical Body (Article 49) whose task will, among other, be to issue recommendations for States Parties planning activities in ABNJ.

In the international legal order, including the above-mentioned instruments, *States* are generally responsible both for the adoption of legal standards for GE and for ensuring that natural and juridical persons under their jurisdiction comply with the required standards and norms when performing GE experiments and projects. Recently, the international legal order has opened up for attempts to transcend State sovereignty in order to protect the international community at large: e.g., collective rights and interests of non-State stakeholders, including indigenous and local communities and future generations. This is demonstrated by advisory proceedings in climate cases before ICJ and ITLOS.[25] The Inter-American Court of Human Rights (IACHR) and European Court of Human Rights (ECtHR) have also dealt with cases addressing legal obligations of States (and public authorities) with regard to climate change.[26]

The human rights based approach is also advocated for by the UN Human Rights Council (UNHRC) as new climate protection technologies may interfere with the enjoyment of human rights, including the right to life, the right to a good environment and to access to justice and effective remedies (UNHRC, 2023).[27] Among other aspects, public awareness and participation is key to protect human rights against harmful impact of technologies, and transparency of GE-related decisions should be one of the central criteria for the future GE assessment framework.[28] These considerations are especially important with regard to the special needs of developing countries that are particularly vulnerable to climate change. Indeed, while GE projects are largely conceptualized in western countries, experiments are reportedly conducted in the southern hemisphere (Chalmin, 2023).

By elucidating international responsibilities of States to protect affected and vulnerable States, communities, and individuals from the harmful impact of climate change, rulings of international courts will have important legal implications for the regulation and governance of GE.[29] ITLOS noted that marine geoengineering may be contrary to obligations envisaged in Article 195 UNCLOS if these activities result in transforming one type of pollution into another. Furthermore, States are under obligation to take all necessary measures to prevent, reduce, and control pollution resulting from the use of technologies under their jurisdiction and control.[30]

Strengthening non-State-centric influence in the international law-making for GE is, in this author's view, generally positive. At the same time, it is doubtful that international judiciary and non-State stakeholders are well-suited to deal with such emergency measures as GE. GE projects which ignore the existence of a treaty framework such as the Antarctic Treaty may result in an 'international discord' (Flamm & Shibata, 2024). If States and international organizations continue to be passive spectators, rather than pro-active international law makers for GE, the international law on this unique and critical issue will still evolve, but in a less predictable way.

4 Exploring Future Research on GE in International Law

The fundamental question for future legal research—recognizable from endeavors to internationally govern other transboundary and novel human activities—is how to regulate and govern GE *effectively*. GE encompasses novel, contentious, and technology-intensive activities, and our insufficient understanding of the goals we seek to achieve hinders the development and adoption of effective regulatory solutions.

One of the persistent and contested issues pertains to determining legally (un)acceptable levels of environmental harm (i.e., legally significant thresholds of 'substantial' or 'significant' harm) and identifying *cumulative* impacts[31] on the environment. This is also the case for GE activities, as it will be challenging to assess ecological impact of particular GE projects in light of other activities, including climate mitigation and adaptation measures. This means that we need to better understand how legal norms and instruments may enable legal communities (international and national alike) to incorporate knowledge acquired from scientific research and to regulate GE on the basis of the best available science and scientific information.

Further, a central task for legal scholars, in this author's view, is to explore what international legal approach (or approaches) may be suitable to achieve the desired goals within GE. The suitability of a measure can be determined by its capacity to adequately stimulate sustainable behavior by the actors involved in GE projects. This raises questions about the contents, shape, and normative effects of future legal instruments, e.g., full or partial prohibition, criteria and standards, 'hard' or 'soft' law measures or a combination of both. Considering that GE techniques will be designed and performed not only by scientists but, quite likely, by commercial actors, it will be necessary to develop rules which stimulate innovative solutions without compromising the relevant international law principles, including the precautionary principle.

The following criteria for a future legal framework may generally be discerned from the work of international legal actors such as IMO, CBD and ILC: (1) legally binding (as proposed by IMO); (2) science-based, (3) global, (4) transparent, and (5) subject to effective regulatory control mechanisms. This inspires several fundamental questions: what kind of legal instrument on GE is realistic and desirable for the international community to work for? Should this be a new, *sui generis* and *binding* instrument on GE, what other options are available—and perhaps more feasible—for international law makers to resort to?

Some scholars have proposed to adopt an international 'no-use' agreement (on SRM) that would combine prohibitions and restrictions of 'specific harmful activities or technologies without placing undue restrictions on legitimate and desirable activities' (Gupta et al., 2024). They reasonably argue that, to develop such novel regime(s) for SRM, it may be necessary to apply legal techniques familiar from other multilateral regimes while also taking account of the unique characteristics of each GE approaches.

Law-making efforts in other areas of international law such as cyber security or outer space show that, in the absence of a competent sectoral organization or another robust institutional framework, it may be impossible to move beyond non-binding, 'soft' norms (Kravik, 2021). However, the binding character of commitments is not a sufficient or appropriate criterion to assess the effectiveness of an instrument in question. For GE, it could, for example, be more decisive to direct norm-creating efforts to the discussion and development of more specific due diligence standards and criteria.

Considering the complexity and fragmentation of international law on GE, presently governed by various sectoral instruments as well as by general provisions of international law, it is also crucial to find solutions which ensure coherence across the complex and fragmented regulatory landscape of sectoral and universal provisions on GE. As discussed earlier, BBNJ provides for a promising 'bridge' between various global and regional instruments and provisions, at least with regard to the protection of oceans against harmful impacts of GE. However, the fragmentation and coherence of international law in the context of GE is largely an unplowed field in legal research.

Last but not least, it remains open, by *whom* legal norms and standards for GE will (and should) be adopted. Judging by experience in other technology-intensive and dynamic sectors, it is predictable that commercial and other non-State stakeholders will develop their own due diligence standards for GE to fill the gaps in the public regulation. Commercial actors may also use this opportunity to emancipate from slow, ineffective, and perhaps burdensome public laws by developing self-regulation (Freeland et al., 2018). While vast literature on legal pluralism already exists in other fields (Santos, 2020), there is need for further research on how this will evolve in the field of GE, including transnational law perspectives and the issue of how such pluralistic approaches to GE could be adequately incorporated in the future GE governance.

5 Conclusions

This chapter argues that internationally harmonized—if not fully uniform—standards for GE are indispensable for the effective governance of these activities. Admittedly, national measures are required for the implementation and enforcement of international norms, especially those regulating commercial and other non-state actors. However, domestic regulation is unlikely to be successful in the absence of some harmonized criteria for GE, due to the transboundary impact of GE and the likely differences among stakeholders in how they perceive risks of GE (UNEP, 2023). In the absence of a uniform approach at the international level, states may allow (or may not be able to preclude) large-scale GE experiments and, potentially, give a green light to commercial development and application of SRM projects without sufficiently considering their risks, uncertainties, and impacts (UNEP, 2023).

While a prohibition on GE has been advocated by some States and organizations, it is, in this author's view, feasible to focus efforts on the development of effective, workable norms and procedures to govern GE. Examples of CBD and IMO show that a (functional) prohibition may be accomplished by the adoption of criteria and procedures to govern GE projects, to the extent they are governed by these frameworks. However, the discussion in this chapter shows that key issues of international law and governance of GE remain debatable, including the desirability of a universal, international initiative to address GE in multilateral instruments. This gives reasons to fear that sectoral accomplishments of the international community under auspices of CBD and IMO will turn out to be exceptions, not followed up by international regulatory attempts in the UN or other multilateral frameworks.

To conclude, the international community has not yet used the full potential of international law to tackle emerging GE issues. It remains to be seen how the gaps in the international regulation of GE will be filled: by public international law measures adopted by States and international organizations or rather by self-regulation and judicial law-making. While the latter two approaches may be a welcome addition to the GE governance framework, they are poorly suited to deal with emergency or crises situations that GE has an ambition to solve. Lingering in the development of international norms on GE will not hinder technological and scientific progress and, ultimately, commercial deployment of GE techniques. The lack of harmonized approach to GE may amplify the risk that individual States will authorize GE projects in contravention of environmental law principles, including the precautionary principle. In the next decades, the legal research community, in collaboration with other sciences, faces challenging tasks, as it has to promote knowledge-based decision-making on GE at the international and national level and to work out feasible and effective legal solutions for GE.

Notes

1. See chapters "" and "SWOT Analysis for Solar Radiation Modification (SRM)" on SWOTs for SRM and CDR.
2. See chapter "Introduction: Turning Earth's Thermostat Down—But How? The Politics ofGeoengineering. Perspectives from the Social Sciences" Introduction.
3. See chapter "Who Will Bell the Cat? African Union, Geoengineering and the Potential Role of Human Rights" on human rights and this chapter on ecocide.
4. Convention on Biological Diversity, Rio de Janeiro, 5 June 1992, in force 29 December 1993, 1760 UNTS 69 COP Decision X/33. https://www.cbd.int/decision/cop/?id=12299.
5. See also chapter "Governing Carbon Dioxide Removal and Solar Radiation Modification" on international relations and governance of GE.

6. CBD COP 9 Decision IX/16. (2008). Biodiversity and Climate Change. https://www.cbd.int/doc/decisions/cop-09/cop-09-dec-16-en.pdf.
7. COP 10 Decision X/33.(2010). Biodiversity and Climate Change. https://www.cbd.int/decision/cop/?id=12299.
8. Convention on the Prevention of Marine Pollution by Dumping of Wastes and Other Matter, Washington/Moscow/London/Mexico City, adopted 29 December 1972, in force 30 August 1975, 1046 UNTS 120.
9. 1996 Protocol to the Convention on the Prevention of Marine Pollution by Dumping of Wastes and Other Matter, 1972 (amended in 2006).
10. Annex I to LC (n 8).
11. Resolution LC-LP.1. (2008). On the Regulation of Ocean Fertilization. https://wwwcdn.imo.org/localresources/en/KnowledgeCentre/IndexofIMOResolutions/LCLPDocuments/LC-LP.1%20(2008).pdf. Annex 4 added to LP by the Resolution LP.4(8), ibid.
12. Resolution LC-LP.2. (2010). On the Assessment Framework for Scientific Research Involving Ocean Fertilization. https://wwwcdn.imo.org/localresources/en/KnowledgeCentre/IndexofIMOResolutions/LCLPDocuments/LC-LP.2(2010).pdf.
13. Resolution LP.4(8). (2013).On the Amendment to the London Protocol to Regulate the Placement of Matter for Ocean Fertilization and Other Marine Geoengineering Activities, Article 6bis. Not in force. https://wwwcdn.imo.org/localresources/en/KnowledgeCentre/IndexofIMOResolutions/LCLPDocuments/LP.4(8).pdf.
14. Article 1 paragraph 5bis of LP, added by Resolution LP.4(8) (ibid.). (2013). On the Amendment to the London Protocol to Regulate the Placement of Matter for Ocean Fertilization and Other Marine Geoengineering Activities.
15. IMO. (2022). Marine geoengineering techniques - potential impacts. https://www.imo.org/en/MediaCentre/PressBriefings/pages/Marine-geoengineering.aspx.
16. Montreal, 16 September 1987, 1522 UNTS 3, in force 1 January 1989.
17. Vienna, 22 March 1985, 1513 UNTS 293, in force 22 September 1988.
18. International Law Commission. (2021.) Draft Guidelines on the protection of the atmosphere. *Yearbook*, II(2), Guideline 7.
19. Guideline 2(2) (ibid.).
20. See chapter Governing Carbon Dioxide Removal and Solar Radiation Modification and this chapter.
21. UNGA Resolution 1721. (XVI)(1961). International cooperation on peaceful uses of outer space, C 1(a).
22. Some SRM techniques may fall within the scope of 'space activities' governed by the Treaty on Principles Governing the Activities of States in the Exploration and Use of Outer Space, including the Moon and other celestial bodies (the Outer Space Treaty) London/Moscow/Washington, 19 December 1966, in force 10 October 1967, 610 UNTS 205.
23. As Gupta (2016) points out, the scientific community is not united in its views on GE; to the contrary, significant disagreement exists between scientists.
24. New York, 19 June 2023, not in force. The text of the Agreement is available at https://www.un.org/bbnj/.
25. ITLOS. (2024, Advisory Opinion of 21 May 2024, Case No. 31). ICJ. (2025, Advisory Opinion of 23 July 2025).
26. IACHR (November 15, 2017). Case OC-23/17 Advisory Opinion. The Environment and Human Rights. https://climatecasechart.com/wp-content/uploads/non-us-case-documents/2017/20171115_OC-2317_opinion-4.pdf (English translation); IACHR (May 29, 2025). Case AO-32/25 Advisory Opinion; ECtHR (2024). Case of Verein Klimaseniorinnen Schweiz and Othres v. Switzerland (Judgment, Application no. 53600/20).
27. See chapter "Who Will Bell the Cat? African Union, Geoengineering and the Potential Role of Human Rights".
28. See also IMO (2022) Background information on LP/LC ocean fertilization resolutions and the 2013 LP amendment (Marine geoengineering techniques - potential impacts [https://www.imo.org/en/MediaCentre/PressBriefings/pages/Marine-geoengineering.aspx]).

29. UNGA Resolution 77/726 of 29 March 2023 Request for an advisory opinion of the International Court of Justice on the obligations of States in respect of climate change. UNGA also highlights the special needs of developing countries that are 'particularly vulnerable' to these effects.
30. ITLOS (n 25) para 231.
31. Article 1(6) of BBNJ (n 24) defines 'cumulative impacts' as 'the combined and incremental impacts resulting from different activities, including known past and present and reasonably foreseeable activities, or from the repetition of similar activities over time, and the consequences of climate change, ocean acidification and related impacts'.

References

Bohman, B., & Ringbom, H. (2023). Marine geoengineering to abate eutrophication in the Baltic Sea: How to address regulatory voids and uncertainty. In F. M. Platjouw & A. Pozdnakova (Eds.), *Environmental rule of law for oceans: Designing legal solutions* (pp. 108–122). Cambridge University Press. https://doi.org/10.1017/9781009253741.013

Brent, K. A. (2021). Solar geoengineering is prohibited under international law. In B. Mayer & A. Zahar (Eds.), *Debating climate law* (pp. 274–284). Cambridge University Press. https://doi.org/10.1017/9781108879064.021

Brunnée, J. (2019). The rule of international (environmental) law and complex problems. In H. Krieger, G. Nolte, & A. Zimmermann (Eds.), *The international rule of law: Rise or decline?* (pp. 211–236). Oxford University Press. https://doi.org/10.1093/oso/9780198843603.003.0014

Chalmin, A. (2023, February 15). *Commercialization in geoengineering continues to increase*. Geoengineering Monitor. Retrieved February 20, 2025, from https://www.geoengineeringmonitor.org/2023/02/quarterly-review-iv/

COPUOS. (2024). *Space for sustainable development: Technology and its applications, including the United Nations Programme on Space Applications*. Retrieved January 15, 2025, from https://www.unoosa.org/res/oosadoc/data/documents/2024/aac_105c_1l/aac_105c_1l_411add_7_0_html/AC105_C1_L411Add07E.pdf

Corbett, C. R. (2022). The climate emergency and solar geoengineering. *Harvard Environmental Law Review, 46*(1), 197–260.

Flamm, P., & Shibata, A. (2024). 'Ice sheet conservation' and international discord: Governing (potential) glacial geoengineering in Antarctica. *International Affairs*, iiae281. https://doi.org/10.1093/ia/iiae281

Freeland, S., Hutchison, K., & Sim, V. (2018). How technology drives space law down under: The Australian and New Zealand experience. *Air and Space Law, 43*(2), 129–147. https://doi.org/10.54648/aila2018010

GESAMP. (2019). *High level review of a wide range of proposed marine geoengineering techniques*. Retrieved December 10, 2024, from http://www.gesamp.org/publications/high-level-review-of-a-wide-range-of-proposed-marine-geoengineering-techniques

Gupta, H. (2016). Geoengineering: Issues and perspectives. *Environmental Policy and Law, 46*(1), 50–55. https://doi.org/10.3233/EPL-46103

Gupta, A., Biermann, F., van Driel, E., Bernaz, N., Jayaram, D., Kim, R. E., Kotzé, L. J., Ruddigkeit, D., VanDeveer, S. D., & Wewerinke-Singh, M. (2024). Towards a non-use regime on solar geoengineering: Lessons from international law and governance. *Transnational Environmental Law*, 1–32. https://doi.org/10.1017/S2047102524000050

IPCC. (2011). *IPCC expert meeting on geoengineering*. Retrieved November 5, 2024, from https://archive.ipcc.ch/pdf/supporting-material/EM_GeoE_Meeting_Report_final.pdf

IPCC. (2022). *Climate change 2022: Mitigation of climate change. Contribution of Working Group III to the Sixth Assessment Report of the Intergovernmental Panel on Climate Change*. Cambridge University Press. https://doi.org/10.1017/9781009157926

Kravik, A. M. (2021). An analysis of stagnation of multilateral law-making—And why the law of the sea has transcended the stagnation trend. *Leiden Journal of International Law, 34*(4), 935–956. https://doi.org/10.1017/S092215652100039X

Lee, J. (2011, February 23). *Eastern Pacific Emitted Aerosol Cloud Experiment (E-PEACE)*. ClimateViewerNews. Retrieved January 10, 2025, from https://climateviewer.com/eastern-pacific-emitted-aerosol-cloud-experiment-e-peace/

Lerner, K. L. (2021). Geoengineering. In K. H. Nemeh & J. L. Longe (Eds.), *The Gale encyclopedia of science* (pp. 1973–1975). Gale Academic.

Low, S., Moore, N., Chen, Z., McManamen, K., & Blackstock, J. J. (2013). Geoengineering policy and governance issues. In T. Lenton & N. Vaughan (Eds.), *Geoengineering responses to climate change: Selected entries from the encyclopedia of sustainability science and technology* (pp. 169–191). Springer. https://doi.org/10.1007/978-1-4614-5770-1

Mayer, B. (2018). *The international law on climate change*. Cambridge University Press.

Morgera, E. (2011). Far away, so close: A legal analysis of the increasing interactions between the Convention on Biological Diversity and climate change law. *Climate Law, 2*(1), 85–115. https://doi.org/10.1163/CL-2011-027

Njagi, D. (2024, March 1). *Solar geoengineering rejected at 6th UN Environment Assembly*. Devex. Retrieved March 5, 2024, from https://www.devex.com/news/solar-geoengineering-rejected-at-6th-un-environment-assembly-107157

Proelss, A. (2012). Geoengineering and international law. *Security and Peace, 30*(4), 205–211. https://doi.org/10.5771/0175-274x-2012-4-205

Royal Society. (2024). *UK Norway bilateral workshop on science and the law of the environment meeting report*. Retrieved January 20, 2025, from https://royalsociety.org/-/media/policy/publications/2024/des8768_1_uk-norway-bim_conference-report_web.pdf

Santos, B. de S. (2020). *Toward a new legal common sense: Law, globalization, and emancipation*. Cambridge University Press. https://doi.org/10.1017/9781316662427

Scott, K. N. (2023). From ocean dumping to marine geoengineering: The evolution of the London regime. In R. Rayfuse, A. Jaeckel, & N. Klein (Eds.), *Research handbook on international environmental law* (pp. 240–263). Edward Elgar. https://doi.org/10.4337/9781789909081

Steenkamp, R. C. (2025). *International law and marine geoengineering*. Nomos. https://doi.org/10.5771/9783748953111

UARCTIC. (2024, May 2). *Seabed curtain project*. Retrieved May 2, 2024, from https://www.uarctic.org/activities/thematic-networks/frozen-arctic-conservation/seabed-curtain-project/

UK-DNVA. (2023). *UK-Norway bilateral workshop on science and the law of the environment*. The Royal Society.

UNEP. (2023). *One atmosphere: An independent expert review on solar radiation modification research and deployment*. Retrieved December 15, 2024, from https://wedocs.unep.org/bitstream/handle/20.500.11822/41903/one_atmosphere.pdf?sequence=3&isAllowed=y

UNHRC. (2023). *Impact of new technologies intended for climate protection on the enjoyment of human rights* (A/HRC/54/47). Retrieved January 12, 2025, from https://documents.un.org/doc/undoc/gen/g23/141/86/pdf/g2314186.pdf?token=oJGW96trWqlepL8F5z&fe=true

Voigt, C. (2023). Oceans and climate change: Implications for UNCLOS and the UN climate regime. In F. M. Platjouw & A. Pozdnakova (Eds.), *Environmental rule of law for oceans: Designing legal solutions* (pp. 17–30). Cambridge University Press. https://doi.org/10.1017/9781009253741.006

Watts, J. (2024, February 22). Switzerland calls on UN to explore possibility of solar geoengineering. *The Guardian*. Retrieved February 25, 2024, from https://www.theguardian.com/environment/2024/feb/22/switzerland-calls-on-un-to-explore-possibility-of-solar-geoengineering

Solar Geoengineering and the Global Commons—What Role for Ecocide Law?

Anna Faye Maddrick

Abstract This paper assesses the concepts of solar geoengineering and ecocide law, arguing that while wide-scale instantiation of both solar geoengineering deployment and the implementation of a crime of ecocide remain, as yet, theoretical, there exist many obvious parallels between the two phenomena. Firstly, by briefly reviewing definitional and historical parallels between the concepts, including their focus on scale and intent, it will be argued that ecocide law could naturally build upon existing international law and serve as a useful analytical tool for reviewing the legality or otherwise of solar geoengineering deployment. Secondly, it is argued that ecocide law, in conjunction with polycentric governance, could ensure more adequate environmental protection for the global commons in the face of solar geoengineering, as well as providing a legitimate and robust legal framework for discussing any such developments, reinforced by internationally enforceable rights on planet earth and in outer space.

Keywords Ecocide law · Solar geoengineering · Global commons · Judicial enforcement · Outer space · Polycentric governance

1 Introduction

> This view of climate history is in accord with current understanding of the history of science that sees the drive to manipulate nature to suit human ends as integral to the process by which knowledge is accumulated. In this view, the drive to impose human rationality on the disorder of nature by technological means constitutes a central element of the modernist program. (Keith, 2002, p. 79)

Ecocide, according to the most authoritative definition, now proposed as an amendment to the Rome Statute, is understood as "*unlawful or wanton acts committed with*

A. F. Maddrick (✉)
University of Bologna, Bologna, Italy
e-mail: maddrickanna@gmail.com

Permanent Mission of Vanuatu to the United Nations, New York, NY, USA

K.-U. Schrogl et al. (eds.), *The Politics of Geoengineering*, Springer Climate,
https://doi.org/10.1007/978-3-032-10095-5_9

knowledge that there is a substantial likelihood of severe and either widespread or long-term damage to the environment caused by those acts" (Independent Expert Panel [IEP], 2021). Etymologically derived from the Greek prefix "oikos", meaning house, and the Latin suffix "cide" for caedere, i.e. to kill, ecocide is the act of destroying our collective home, understood as Planet Earth, and at the root of the growing movement for the international criminalisation of ecocide law is the protection of the earth/biosphere as the common good of humanity (Vishwas, 2018, p. 54), which requires the necessary interventions to be taken in order to stop and avert significant dangers for present and future generations. In order to propose a comprehensive and inclusive offence that reaches a wide variety of actors and can therefore effectively punish behaviour, as well as reflect the environment's natural interconnections,[1] an international crime of ecocide is based in this definition on the five main spheres of the earth (Voigt, 2021), thus defined as "*the earth, its biosphere, cryosphere, lithosphere, hydrosphere and atmosphere, as well as outer space*" (IEP, 2021).

Climate change is widely considered the greatest "collective action" problem of history, and as the world's most vulnerable populations suffer from sharply rising temperatures, global decision-makers are under increasing pressure to act. Some have argued that such pressure will apply to geoengineering deployment, due to its allegedly cheaper and more rapid effects than other mechanisms. Geoengineering, for purposes here[2] defined as "*the intentional large-scale manipulation of the environment, particularly manipulation that is intended to reduce undesired anthropogenic climate change*" (Keith, 2002, p. 245), is increasingly entering international fora as a potential strategy for dealing with current and future climate change. Admittedly, what classifies as "significant" or large-scale, and thus able to meet the definition of geoengineering, is an open-ended judgement (McMartin et al., 2016). However, wherever the line is drawn, it will exclude research activities having no or low physical impacts on the environment (Corbett, 2022, p. 235). Discussion on geoengineering often differentiates between two categories of technologies, namely Carbon Dioxide Removal (CDR) and Solar Radiation Management or Solar Radiation Modification (SRM) and the IPCC adopts this distinction (IPCC, 2011).

Given each practice poses its own particular regulatory challenges (Cook & Sands, 2021), and CDR has been covered by other contributions within this book, I restrict my discussion to SRM, referring to various technologies aimed at reducing incoming sunlight to the earth. SRM deployment poses a unique cost–benefit analysis legally and in our relationship with nature, and as a potential geoengineering measure is well understood to be the most controversial and potentially far-reaching in consequence (Cook & Sands, 2021; IPCC). The deliberate management of the skies on a global scale would, at least in part, force us to view the biosphere as an artefact, thus blurring a sharp distinction between the natural and the artificial, between humanity and our technology (Keith, 2002, pp. 277–278).

While both wide-scale use of SRM and the implementation of an international crime of ecocide, as of yet, remain theoretical, in this contribution I consider that obvious parallels exist between the two definitions. Scale and intent play a central role in the definitions of both geoengineering (IPCC, 2011; Keith, 2002, p. 247) and ecocide law. As Keith notes, "*three core attributes will serve as markers of geoengineering: scale, intent and the degree to which the action is a countervailing measure*" (Keith, 2002, p. 247). Similarly, whether an act or omission is constitutive of ecocide depends on the scale of environmental damage as severe and either widespread or long-term; the "*knowledge of a substantial likelihood*"; and whether such actions were "*unlawful or wanton*" (IEP, 2021).

Industrialisation and technological advancement have given a small proportion of individuals influence over the planet that is "*vast, clumsy, poorly understood, and often difficult to predict*" (Corbett, 2022, p. 241). SRM in particular threatens a safe transition away from destructive structures and practices by entrenching the exact systems that need redesigning (Hands Off Mother Earth, 2018), posing enormous unknown risks. Intent and scale therefore distinguish SRM from other human geoengineering activities with planetary-scale impacts, such as use of nitrogen-based fertilisers or even fossil fuel energy systems (Corbett, 2022). As this discussion will focus on intent and scale, I will use SRM for analysis but consider that these arguments could be applicable to any wide-scale use of geoengineering practices. A regime of active planetary climate modification, even if eventually necessary,[3] would strain existing systems of international law and demand new institutions without clear precedent, being marked by high levels of scientific uncertainty. More intuitively, such a regime provokes disconcerting questions about the interactions between humans, the earth, and outer space. SRM as a climate mitigation measure is, by design, rooted in the same philosophy that is responsible for SRM's very existence as a proposed solution. Consequently, I take the view that "[h]*umanity may inevitably grow into active planetary management, yet we would be wise to begin with a renewed commitment to reduce our interference in natural systems rather than to act by balancing one interference with another*" (Keith, 2002, p. 280).

In this contribution, I will firstly discuss ecocide law's potential *direct* relevance to SRM deployment by definitional comparison, with references to historical precedent. I will argue that ecocide law is well reflective of existing international law in this area, revealing some theoretical considerations. Thereafter, by reference to principles and application of international space law, and ideas around regulation of the global commons more generally, I will highlight ecocide law's *indirect* relevance to SRM deployment, arguing in favour of polycentric governance supplemented by ecocide law for effective environmental protection against SRM deployment, as well as the provision of legitimate and robust fora for discussing any such developments, reinforced by internationally enforceable environmental and human rights on earth and in outer space.

2 Definitional and Historical Parallels

While geoengineering as a potential mechanism for dealing with climate change is relatively novel, humankind's intentional manipulation of the weather holds a strong history—both for benign[4] and warfare purposes (*Duke Law Journal*, 1960; Hough,

2014; Kolpak, 2020; Munoz, 2017). The historical interlinkages between weather modification, international warfare, and environmental protection bear relevance to current discussions around geoengineering as a deliberate climate modification action. As with other geoengineering proposals, deliberate and inadvertent climate modification are closely linked. Addition of aerosol to the stratosphere could have serious impacts, most notably, depletion of stratospheric ozone (Keith, 2002, p. 262). There exist many uncertainties regarding the consequences—and possible risks—of implementing geoengineering solutions at scale. Once employed, their precise impact on natural systems, both regional and global, is unknown. Equally worryingly, some have argued that the consequences of implementing solar radiation management solutions may be perceived as acts of war, which may contribute to existing conflicts or provoke new ones (Kolpak, 2020).

Ecocide law has a historical basis in a time when environmental warfare and weather modification were highly related. In the 1960s, in response to concerns that environmental warfare in the Vietnam War could be constituted by weather modification techniques such as the creation of earthquakes, tsunamis, or cyclones (Colgan & Colgan, 2023; Kolpak, 2020), the international community, including the Soviets and the Americans, cooperated in formulating a draft for what would become the 1976 Convention on the Prohibition of Military or any Hostile Use of Environmental Modification Techniques (ENMOD Convention 1977) (Hough, 2014, p. 146). After the United States began using herbicides to further its war efforts, scientists quickly retaliated against its usage and claimed that the widespread and indiscriminate herbicide use on jungles and rice paddies throughout Southeast Asia were clear violations of well-founded rules of armed conflict, which led to the emergence of the term ecocide (Cook et al., 1970; Westing, 1974). Consequently, the emergence of ecocide as a term directly stems from the United States' military use of Agent Orange and the immediate outcry that followed, in a time where there was international concern that the US strategy in Vietnam could evolve to include tactics such as weather manipulation; this did much to inspire ENMOD in the first place (Bingle,2022; Hough, 2014).

The environment is an important focus in relation to the broader notion and analysis of threats to international peace and security (Hough, 2016, p. 10; International Criminal Court, 2012). Collateral environmental damage has long been regarded as an inevitable consequence of armed conflicts. But there have also been many examples of the *intentional* manipulation of the environment by warring parties, in which the environment becomes a weapon of war, such as scorched earth (and environmental infrastructure) and utilising the environment in conflict (Kolpak, 2020). Such devastation has often been justified on the grounds of military necessity in ways that could not so readily be done if the casualties were human. Today, military ecocide is prohibited widely under international customary and treaty law; however, historical examples of where the environment, and environmentally risky infrastructure, have been used as a weapon of war are commonplace. Therefore, the potential weaponisation of geoengineering technologies cannot be dismissed out of hand, and any move to develop or deploy them must take these serious risks into account (Kolpak, 2020).

The ENMOD Convention applies to environmental modification techniques that are simultaneously (a) hostile and (b) cause destruction, damage, or injury at, or above, the set threshold. As a result, initially peaceful, yet ultimately harmful geoengineering projects could exemplify unintentional hostile use of weather modification technologies. However, ENMOD does not take this form of use into account, so cannot provide a viable tool for regulating future use of such technologies (Colgan & Colgan, 2023). Moreover, ENMOD, like other law regulating military ecocide, leaves the obvious gap that any harm to the environment is considered secondary to human or economic benefit.

Ecocide law explicitly builds upon this framework. Borrowing from ENMOD and Protocol 1 of the Geneva Convention, to constitute seriousness for a crime of ecocide, the impact(s) must be "*severe and either widespread or long-term*"[5]: with severe as "*damage which involves very serious adverse changes, disruption or harm to any element of the environment, including grave impacts on human life or natural, cultural or economic resources*",[6] widespread as "*damage which extends beyond a limited geographic area, crosses state boundaries, or is suffered by an entire ecosystem or species or a large number of human beings*",[7] and long-term as "*damage which is irreversible or which cannot be redressed through natural recovery within a reasonable period of time*"[8] (IEP, 2021) In any event, the "*knowledge of substantial likelihood*" will be met where it is evident that the damage is likely to be irreversible and carry long-term effects, or unable to be redressed in a reasonable period of time (IEP Commentary, 2021). The second threshold of such actions or omissions being "*unlawful or wanton*" ensures that legitimate actions causing damage to the environment, such as in pursuit of development, are not caught by the scope of the crime unless disproportionately severe.

Culpability for the crime of ecocide attaches to the creation of a dangerous situation, rather than to a particular outcome. It is the commission of acts with knowledge of the substantial likelihood that they will cause severe and either widespread or long-term damage that is criminalised. The crime of ecocide is thus formulated as a crime of endangerment rather than of material result.[9] This avoids the problems of having to await materialisation of ecological catastrophe (Robinson, 2022, p. 332). Endangerment is also key to Articles 35(3) and 55(1) API, which prohibit the use of "*methods or means of warfare which are intended, or may be expected to cause, widespread, long-term and severe damage to the natural environment*" (Steffen et al., 2020). As a result, ecocide law could fill a gap in the international legal framework: ensuring environmental damage is not condemned solely because it is a consequence of warfare, or committed for hostile purposes, rather it can also be recognised and punished during peacetime, and potentially due to a wrong against the environment itself without immediate human harm.

Due to significant scientific uncertainties, any present deployment of SRM is going to be committed with knowledge of a risk of severe and either widespread or long-term damage to the environment. The technology is already subject to a wide spectrum of international outcry (Alaska Native Delegation; Don't Geoengineer Africa; Solar Geoengineering Non-Use Agreement, 2022), on the understanding that any deployment at planetary-scale cannot be fairly and effectively governed in the current system of international institutions[10] (Solar Geoengineering Non-Use Agreement, 2022). On its face, therefore, *unilateral* deployment of solar geoengineering at present would be constitutive of ecocide liability. Moreover, even if an internationally lawful solar geoengineering regime could be implemented, any deployment may still be constitutive of ecocide if determined to be committed "*wantonly*". "*Wanton*" means with reckless disregard for damage which would be clearly excessive in relation to the social and economic benefits anticipated (IEP Commentary, 2021). As Keith highlights, even if strong arguments can be levied that a geoengineering practice is a socially optimal strategy, it may still present a moral hazard if its implementation encourages a retreat from agreed stringent action on mitigation (Keith, 2002, p. 276), which protestors against SRM repeatedly affirm (Hands off Mother Earth). Similarly, geoengineering may also create environmental security risks in third countries, where poorer countries lack the resources to participate in development and deployment, but are still subject to the technology's inevitably transboundary consequences, increasing potential for conflict (Boyd, 2009; Kolpak, 2020). As a result, it is well argued that the ecocide law "*wanton*" standard could provide a useful conceptual tool for governing solar geoengineering deployment.

This calculation would ensure that any such deployment is a legally buffered method of *absolute* last resort and would avoid the potential scenario noted by Cook and Sands in which one might invoke the precautionary principle *in favour* of SRM if the world is on a path to exceed a 3 °C temperature increase, in itself constituting a likely risk of serious harm (Cook & Sands, 2021, p. 27) and implying that failure to deploy SRM in the face of climate change effects would be constitutive of ecocide by omission (turning my argument on its head). It is, after all, hard to see what social and economic benefits could be anticipated by allowing dangerous climate change to unfold upon humanity. This scenario, however, falls prey to a logical fallacy all too often employed by those with vested interests in particular solutions: just because a given situation presents a high risk does not in any way imply that any one technology or approach must be favoured over others to counteract that risk.[11]

Indeed, SRM is a highly uncertain technology with likely irreversible consequences, posing a novel range of ethical and regulatory challenges (McMartin et al., 2016) and serious practical risks. Quite apart from the unpredictability of complex effects on the biosphere and on humanity, it well may detract from meaningful mitigation measures or even lock in dependence on fossil fuels (Cook & Sands, 2021). As a result, especially given the international community's continuing failure to deliver basic commitments on climate change, it can be well argued that recourse to SRM at this stage would be wanton and thus instantly constitutive of an ecocide crime. As outlined above, SRM also straddles a disconcerting interface between warfare and the environment. Given the undoubtable necessity of strengthened international cooperation in environmental protection, as well as the existing inadequacies of international environmental law, we would be wise to pay attention to historical lessons, legislating anticipatively against such serious risks.

3 Ecocide Law and Polycentric Governance of Global Commons

While many nations view space as the "common heritage of mankind", a resource not owned by any nation but crucial to the future of all the international community (Fisk, 2022; United Nations, 1967) a debate around the cornerstone principle of non-appropriation (United Nations, 1967, Article II) is ongoing. Alongside a deficit of regulatory law, there is a significant trend in the space community towards the commoditisation (Rhimbassen & Rapp, 2021) and militarisation (Marshall, 2023) of the space economy and space resources. Outer space is therefore particularly at threat in the areas of peace and arms control, sustainability of use, and just and fair distribution of resources and benefits (Schrogl, 2021, p. 99). SRM deployment is relevant to all three concerns, and consequently, it is argued that States need to coalesce behind a commons-based understanding and legal application to manage present and future risks. Repercussions of the climate and environmental crises are transboundary and trans-generational and the challenge of legal reform is consequently interspatial, global, and inter-temporal. SRM as a "solution" adds a further dimension to this complexity, and it is therefore essential to propose comprehensive, inclusive and adaptive legal frameworks that can reach a wide variety of actors and contexts and preserve rights: for nature and for people. It is considered that ecocide law can be pivotal to this process (Andres, 2021).

In applying common-property regimes, which are traditionally more successful at a local scale, to the management of global environmental commons such as outer space, existing academic literature[12] and increasing practice[13] has repeatedly pointed to the concept of "polycentric governance". A polycentric system is defined by diverse decision-making centres that exercise "*considerable independence to make norms and rules within a specific domain*" (Ostrom, 1999, p. 552), in practice a dispersion of authority by level—as applicable—supplemented by the support of private corporations, voluntary associations, and community-based organisations (McGinnis & Ostrom, 2011, p. 15).

Polycentric governance touches upon a number of themes in this discussion. For example, in a discussion of existing weather modification activities in the United States, Simon (2021, p. 211) highlights how a number of polycentric techniques could lead to more equitable governance on weather modification and geoengineering:

> "*Applying iterative learning processes to monitor and review cloud seeding projects, like scenario planning, could contribute to inform and legitimise weather modification decision-making while advancing atmospheric sciences. To do so, the exploitation of atmospheric water requires flexible governance frameworks that take into account scientific uncertainties and facilitates transitions*", further noting (p. 216) that: "*Legal reforms will be needed to facilitate collaboration between institutions, the scientific community and other stakeholders, including local communities, to combine problem-solving competences*".

Whatever one's position on geoengineering, it is evident that deployment of an SRM program's design—including the amount of cooling desired—would be politically contested around the world. Even in idealised scenarios, physical impacts of SRM would not be evenly distributed worldwide or may be exploited by larger powers. Some countries could receive more benefits from a given design while others could receive less. It is even harder to imagine obtaining the consent necessary to make deployment lawful and legitimate: states are obliged to refrain from substantial interference with the environment within the territory of others (U.N. Conference on the Human Environment, 1972).[14]

A polycentric approach to governance specifically takes into account these multiple side effects of action that can incentivise or deter the actions of various actors engaged in governance, mutually organising direct interactions between different centres which could be particularly useful for discussions around geoengineering governance. This approach ranges from promoting recognition of the dependencies between actors whether local, regional, national, trans- or international, to the deliberate organisation of strategic interactions among actors with different preferences and capabilities (Dorsch & Flachsland, 2017, p. 55). To this end, it is important to enable subsidiarity jurisdictions to effectively express their preferences (Ostrom, 1999, 2010). By example, existing space law reveals some tenets of a successful polycentric framework. Among other things, international project-based cooperation in space according to the CHM principle has been successful at the highest level, even outlasting geopolitical tensions (such as the International Space Station with significant American and Russian cooperation) (Schrogl, 2021, p. 102).

On the one hand, therefore, features of polycentric governance may prove useful for governing geoengineering development and any potential deployment, offering a governance paradigm that explicitly incorporates equitable distribution of power and decision-making (Ostrom, 1972).

On the other hand, however, the absence of clear interpretative responsibilities and thus the fragmentation inherent to polycentric governance can enable more powerful interests to prevail in state-free areas (Rhimbassen & Rapp, 2021), as evidenced by the increasing commoditisation of the space environment by powerful actors (United Nations Office for Outer Space Affairs, 2010). The development and deployment of SRM is likely to significantly exacerbate this trend, in the process creating notable adverse environmental impacts upon the ozone layer, outer space, Antarctica, or the high seas. As a result, I argue here in line with other scholars that regulation of state-free areas such as outer space could be improved by the introduction of agreed thresholds and codes of conduct as well as transparency and confidence building measures (Schrogl, 2021, p. 105), with international organisations and forums available as negotiation platforms, and any diplomatic leadership "*ideally*" combining sustainability and distributive justice (Schrogl, 2021, p. 106). By example, Morin and Richard (2021, p. 569) find that top-down regulations, combined with monitoring and sanction mechanisms, can provide adequate legal protection in space. They note that clear and non-discriminatory rules facilitate convergence towards cooperative behaviour, while enforcement mechanisms dissuade free-riding. The authors assert that a clear and stable legal framework with monitoring and penalty procedures could give a new impetus to resolving the space debris problem and furthering the sustainable use of outer space (Morin & Richard, 2021, p. 568). Consequently, argument can be made that the introduction of ecocide as an international crime, with protection offered to all global commons as a result, could provide a clear and non-discriminatory framework to ensure environmental protection in the face of new technologies like SRM. Ecocide law can well offer at least the tenets of a stable legal architecture with built in monitoring and penalty procedures according to fundamental principles of international law. In this light, it is well noted that green criminologists have outlined how a stronger criminal law response to ecocides could significantly improve environmental protection as well as shift societal perceptions (Haines & Parker, 2017; White & Kramer, 2015).

Ecocide law could also help mainstream a human rights approach, particularly relevant as the potential impacts of SRM's highly uncertain technology on human rights are understood to be plentiful (Solar Geoengineering Non-Use Agreement). Even the most inclusive forms of governance over geoengineering cannot guarantee effective participation of future generations whose right to a safe climate system may be violated by use of these technologies (UN General Assembly, 2019). More particularly, from the perspectives of indigenous peoples and local communities across the world, the emphasis on resources and relations to things that are held as common ownership is critical, where community, kinship relations, and relations towards nature and life are highly intertwined with the idea of commons (Alcorn et al., 2005, p. 32). Thus, the harm to global commons posed by SRM would also pose significant harm to such cultures (Alaska Native Delegation). As Liljeblad (2023) highlights, ecocide law raises and holds scope to adjudicate many intersectional issues, connecting environmental welfare to armed conflict, genocide, capitalistic ideology, and indigenous rights claims. As SRM has the potential to impact upon all these factors, and the ICC's explicit intention is to punish the worst forms of human rights abuse (Rome Statute, 1998), ecocide law holds significant promise in this context.

Existing conventions and implementing agreements indicate that states can reach valuable commitments to manage the Earth's great commons. These principles apply to space, but progress on establishing more encompassing space governance principles, enforcement mechanisms, and dispute resolution procedures hinges on states affirming the CHM principle (Silverstein, 2021). By providing an effective "top-down" solution of exacting and uniform environmental standards predicated on human rights and notions of common heritage, ecocide law holds great potential to supplement existing polycentric governance of global commons, as well as the threats posed by SRM.

4 Conclusion

This paper provided a brief synopsis of some interlinkages between the global commons, environmental protection and geoengineering by discussing the uniquely combinative definition of ecocide law: one which reflects themes in environmental warfare and existing weather modification regulation, as well as the wider principles that should regulate environmental resources most at threat from extensive planetary manipulations.

After evidencing the definitional and historical parallels between ecocide law and geoengineering deployment, I argued that polycentric governance, supplemented by ecocide law, could provide a credible framework for buffering the considerable risks of potential SRM deployment, as well as a legitimate legal context for discussions around the technology, if such a time comes.

We would be wise to well reflect on the potential consequences of further playing with nature. Protecting our planetary boundaries protects our peace and security, and the missing step between our rhetoric and action on such developments is the adoption of enforceable law.

Notes

1. There are five main spheres on Earth, which interact in complex ways: the lithosphere (the interior and surface of Earth); the biosphere, (that part of the planet that can support living things); the hydrosphere (areas covered with water); the atmosphere (an envelope of gas); the cryosphere (ice at the poles and elsewhere); see Steffen, W., Richardson, K., Rockström, J., et al. (2020). *The* emergence and evolution of Earth System Science. *Nature Reviews Earth & Environment, 1*, 54–63. Available at https://www.nature.com/articles/s43017-019-0005-6).
2. There is no universally accepted definition of geo engineering. Keith's definition is commonly used, and I considered it substantive for this discussion.
3. Under many current estimates the world is on track to surpass the internationally agreed "safe limits" contained within the Paris Agreement.
4. For example, more than 50 countries around the world are engaged in cloud-seeding worldwide: Munoz, L. M. P. (2017). Seeding change in weather modification. *Globally Bulletin, 66*(1). Available at https://public.wmo.int/en/resources/bulletin/seeding-change-weather-modificationglobally.
5. Geneva Convention Additional Protocol I Relative to the Protection of Victims of International Armed Conflicts 6 U.S.T. 3316, 75 U.N.T.S. 135 (1949), Articles 35 and 55.
6. The military manuals of a number of countries adopt the same definition of 'severe' when addressing environmental harm. The Panel considers this to be a sufficient harm threshold for the purposes of the crime of ecocide. The Panel substituted 'cultural' resources for 'other assets', to make explicit the cultural value of elements of the environment, particularly to indigenous peoples.
7. The drafters note that the ENMOD and Protocol descriptions of "widespread" were too high, being predicated on an area covering a scale of several hundred or thousand square kilometres. They therefore propose damage to be extended "beyond a limited geographic area", as they would exclude particularly egregious acts of environmental damage that harm thousands of people in a single city, or population area.
8. The Drafters note that "'Long-term' refers to the temporal nature of the damage. The Committee on Disarmament has interpreted the closely related term 'long-lasting', the term used in ENMOD, to mean a period of several months or a season, while the background materials to API interpret 'long-term' as a period of decades. Both interpretations were considered problematic—one likely too short and the other potentially too long—as well as somewhat arbitrary". This standard does not require waiting for the reasonable period of time to elapse before a prosecution can be brought—what will constitute a reasonable period of time will depend on the particular circumstances of any situation.

9. This is the case with a number of crimes in the Rome Statute, notably Article 8(2)(b)(iv), the war crime of intentionally launching an attack in the knowledge that such attack will cause … widespread, long-term and severe damage to the natural environment which would be clearly excessive in relation to the concrete and direct overall military advantage anticipated. A similar construction appears in the other 'attack' war crimes, Articles 8(2)(b)(i)–(iii). Another example is provided by the crime of Genocide. Under Article 6 of the Rome Statute there is no requirement that the protected group be actually destroyed, in whole or in part. The crime consists in carrying out actions intended to reach that goal.
10. I emphasise strongly that solar geoengineering should be a considered, at best, only a residual option. This discussion should certainly not be prioritised over meaningful systemic environmental law reform.
11. Further elaboration of this argument is outside the scope of this paper but I thought it worth raising.
12. See generally the sources cited in this section.
13. E.g. the Sustainable Development Goals framework is largely set up polycentrically.
14. U.N. Conference on the Human Environment. (1972). Declaration on the Human Environment, U.N. Doc. A/CONF.48/14/Rev. 1, 5 princ. 21. ("States have, in accordance with the Charter of the United Nations and the principles of international law... the responsibility to ensure that activities within their jurisdiction or control do not cause damage to the environment of other States or of areas beyond the limits of national jurisdiction".)

References

Alcorn, J. B., et al. (2005). Designing alternative frameworks for conserving biodiversity with communities and local governments: A case from Pando, Bolivia. In L.Merino & J. Robinson, *Managing the commons: Conservation of biodiversity.* Instituto Nacional de Ecologia. Available at https://www.researchgate.net/publication/260945288_Managing_the_Commons_Conservation_of_Biodiversity

Alaska Native Delegation. *Letter to Arctic Ice Project.* Available at https://docs.google.com/document/d/1G5b-MSKyV5cI96lyBV140jhyfE-SaPTEt85aVZC5TjM/edit

Andres, J. (2021). *The international crime of ecocide: Regulating environmental degradation in outer space.* Available at https://www.spacelegalissues.com/the-international-crime-of-ecocide-regulatingenvironmental-degradation-in-outer-space/

Bingle, M. (2022). Codifying ecocide as an international atrocity crime: How amending ecocide into the Rome Statute could provide Vietnamese agent Orange victims access to justice. *Hawaii Law Review, 45*(1), 124–156.

Boyd, P. (2009). Geopolitics of geoengineering. *Nature Geoscience, 2*, 812. https://doi.org/10.1038/ngeo710

Cole, D. H. (2011). From global to polycentric climate governance. *Climate Law, 2.* https://doi.org/10.3233/CL2011-042

Colgan, J. D., & Colgan, W. T. (2023). The UN could have a secret legal weapon to fight climate change. *Foreign Policy.* Available at https://foreignpolicy.com/2023/04/05/united-nations-icj-vanuatu-climatechange-law-loss-damage-enmod/

Cook, K., & Sands, P. (2021). *The restriction of geoengineering under international law.* OHCHR. Available at https://www.ohchr.org/sites/default/files/2022-06/Annex-SubmissionCIEL-ETC-HBF-TWNGeoengineering-Opinion.pdf

Cook, R. E., Haseltine, W., & Galston, A.W. (1970). What have we done to Vietnam? In B. Weisberg (Ed.), *Ecocide in Indochina: The ecology of war*. Canfield Press.

Corbett, C. R. (2022). The climate emergency and solar geoengineering. *Harvard Environmental Law Review, 46*, 197–260. Available at https://ssrn.com/abstract=3830109

Don't Geoengineer Africa. *A warning call from African civil society organisations.* Available at https://globalforestcoalition.org/press-release-from-dont-geoengineer-africa-a-warning-call-from-africancivil-society-organizations/

Dorsch, M. J., & Flachsland, C. (2017). A polycentric approach to global climate governance. *Global Environmental Politics, 17*(2), 45–64.

Duke Law Journal. (1960). Legal remedies for "cloud-seeding" activities: Nuisance or trespass? *2*, 305–309. https://doi.org/10.2307/1370980

Fisk, L. A. (2022). *Space as a global commons.* Available at https://www.unoosa.org/documents/pdf/hlf/1st_hlf_Dubai/Presentations/26.pdf

Geneva Convention Additional Protocol I Relative to the Protection of Victims of International Armed Conflicts 6 U.S.T. 3316, 75 U.N.T.S. 135 (1949), Articles 35 and 55.

Haines, F., & Parker, C. (2017). Moving towards ecological regulation: The role of criminalisation. In C. Holley & C. Shearing (Eds.), *Criminology and the Anthropocene* (1st ed.). Routledge.

Hands Off Mother Earth! (2018). *Manifesto against geoengineering.* Available at https://www.geoengineeringmonitor.org/2018/10/hands-off-mother-earth-manifesto-againstgeoengineering/?print=pdf

Hough, P. (2014). Defending nature: The evolution of the international legal restriction of military ecocide. In G. Ziccardi, *The global community yearbook of international law & jurisprudence* (online ed., p. 1). Oxford Academic. https://doi.org/10.1093/acprof:oso/9780190270513.003.0006

Hough, P. (2016). Trying to end the war on the world: The campaign to proscribe military ecocide. *Global Security: Health, Science and Policy, 1*(1), 1–13. https://doi.org/10.1080/23779497.2016.1208055

Independent Expert Panel. (2021). *Legal definition of ecocide, commentary and core text.* Available at https://static1.squarespace.com/static/5ca2608ab914493c64ef1f6d/t/60d7479cf8e7e5461534dd07/1624721314430/SE+Foundation+Commentary+and+core+text+revised+%281%29.pdf

International Criminal Court. (2012). *ICC President tells World Parliamentary Conference "ICC brings retributive and restorative justice together with the prevention of future crimes".* Available at: https://www.icc-cpi.int/news/icc-president-tells-world-parliamentary-conference-icc-brings-retributive-andrestorative

IPCC Expert Meeting on Geoengineering. (2011). Lima, Peru. Available at https://archive.ipcc.ch/pdf/supporting-material/EM_GeoE_Meeting_Report_final.pdf

Keith, D. (2002). Geoengineering the climate: History and prospects. In G. Watts, *Innovative energy strategies for* CO_2 *stabilization* (p. 79). Cambridge University Press.

Kolpak, G. (2020). *From ENMOD to geoengineering: The environment as a weapon of war.* Conflict and Environment Observatory. Available at https://ceobs.org/from-enmod-to-geoengineering-the-environmentas-a-weapon-of-war/

Liljeblad, J. (2023). *Intersections of ecocide, indigenous struggle, & pro-democracy conflict: Implications of post-coup Myanmar for ecocide discourses.* International Crime of Ecocide: New Perspectives, Symposium. Available at https://ecocidelaw.com/wp-content/uploads/2023/08/9-Liljeblad-Intersections-ofecocide-Indigenous-struggle-and-pro-democracy-conflict.pdf

Marshall, T. (2023). *The future of geography: How power and politics in space will change our world.* Elliot and Thomson Ltd.

McGinnis, M., & Ostrom, E. (2011). Reflections on Vincent Ostrom, public administration and polycentricity. *Public Administration Review., 72*(1), 15–25. https://doi.org/10.1111/j.1540-6210.2011.02488.x

McMartin, D. G., et al. (2016). Geoengineering with stratospheric aerosols: What do we not know after a decade of research? *Earth's Future, 4*, 543–548. https://doi.org/10.1002/2016EF000418

Morin, J. F., & Richard, B. (2021). Astro-environmentalism: Towards a polycentric governance of space debris. *Global Policy, 12*(4), 568–573. https://doi.org/10.1111/1758-5899.12950

Munoz, L. M. P. (2017). Seeding change in weather modification. *Globally Bulletin, 66*(1). Available at https://public.wmo.int/en/resources/bulletin/seeding-change-weather-modification-globally

Ostrom, V. (1972). Polycentricity. In M. McGinnis (Ed.), *Polycentricity and local public economies: Readings from the workshop in political theory and policy analysis*. University of Michigan Press.

Ostrom, E. (1999). Coping with tragedies of the commons. *Annual Review of Political Science, 2*, 493–535. https://doi.org/10.1146/annurev.polisci.2.1.493

Ostrom, E. (2010). Beyond markets and states: Polycentric governance of complex economic systems. *American Economic Review, 100*(3), 641–672.

Rhimbassen, M. L., & Rapp, L. (2021). New space property age: At the crossroads of space commons, commodities and competition. *Journal of Property, Planning and Environmental Law, 13*(2), 88–106.

Robinson, D. (2022). Ecocide—Puzzles and possibilities. *Journal of International Criminal Justice, 20*(2), 313–347.

Schrogl, K. U. (2021). Perspectives for outer space as a 'global common.' In M. Benkoe & K. U. Schrogl (Eds.), *Outer space—Future for humankind. Issues of law and policy* (pp. 93–101). Eleven International Publishing.

Silverstein, B. (2021). *Space is a great commons: It's time to treat it as such.* Carnegie Endowment for International Peace. Available at https://carnegieendowment.org/2021/03/09/space-is-great-commons.-it-s-time-totreat-it-as-such-pub-84018#:~:text=Most%20famously%2C%20the%20Outer%20Space,arise%20in%20subsequent%20international%20texts

Simon, M. (2021). Enhancing the weather: Governance of weather modification activities in the United States. *William & Mary Environmental Law and Policy Review, 46*(1). Available at https://scholarship.law.wm.edu/wmelpr/vol46/iss1/5

Solar Geoengineering Non-Use Agreement. *Open Letter.* Available at https://www.solargeoeng.org

Steffen, W., Richardson, K., Rockström, J., et al. (2020). The emergence and evolution of Earth System Science. *Nature Reviews Earth & Environment, 1*, 54–63. Available at: https://www.nature.com/articles/s43017-019-0005-6

The Convention on the Prohibition of Military or any Other Hostile Use of Environmental Modification Techniques. (1977). 1108 U.N.T.S. 151.

UN General Assembly. (2019). *Human rights obligations relating to the enjoyment of a safe, clean, healthy and sustainable environment.* UN Doc. A/74/161. Available at https://www.ohchr.org/sites/default/files/2022-05/20220527-wewerinke-singh-leiden-universitySolargeoNUA%20.pdf

United Nations. (1967). *Treaty on Principles Governing the Activities of States in the Exploration and Use of Outer Space, including the moon and other celestial bodies.* 18 U.S.T. 2410 610 U.N.T.S. 205, 61 I.L.M. 386.

United Nations. (1998). *Rome Statute of the International Criminal Court.* ISBN No. 92-9227-227-6.

United Nations Office for Outer Space Affairs. (2010). *Space debris mitigation guidelines of the Committee on the Peaceful Uses of Outer Space.* Available at: https://www.unoosa.org/pdf/publications/st_space_49E.pd

Vishwas, S. (2018). The Anthropocene and imperial ecocide: Prospects for just transitions. In V. Satgar (Ed.), *The climate crisis: South African and global democratic eco-socialist alternatives.* Wits University Press. https://doi.org/10.18772/22018020541.8

Voigt, C. (2021). *Ecocide as an international crime: Personal reflections on options and choices.* EJIL:Talk!, Available at: https://www.ejiltalk.org/ecocide-as-an-international-crime-personal-reflections-onoptions-and-choices/#:~:text=The%20architectural%20choice%20fell%20therefore,our%20current%20state%20of%20knowledge (last accessed 03/02/23).

Westing, A. (1974). Proscription of ecocide. *Science and Public Affairs, 26*, 24–27.

White, R., & Kramer, R. (2015). Critical criminology and the struggle against climate change ecocide. *Critical Criminology, 23*(4). https://doi.org/10.1007/s10612-015-9292-5

Geoengineering and Outer Space: Policy and Legal Considerations

Anthi Koskina and Manolis Plionis

Abstract Space-based Geoengineering (GE), understood as the deliberate and benevolent manipulation of planetary environment from or in space may be regarded as a possible option if current mitigation or adaptation measures to Climate Change prove insufficient. It could be categorized into two types, both intended to achieve similar objectives: (a) GE technologies directly aimed at altering the environment's composition, and (b) GE solutions based on the use of novel untested technologies, like space-based solar shields. Should such a path be pursued, their use would require careful legal regulation to ensure compliance with the Outer Space Treaty (1967), especially Art. IX aimed at ensuring protection of the natural environment lato sensu. Nevertheless, to address potential challenges arising from unfamiliarity with this new type of activity, legal experts and decision-makers may need to review legislations and even consider whether new approaches should be applied, rather than a strict, literal reading of the existing International Space Law (ISL).

Keywords Space-based Geoengineering (GE) · Carbon Dioxide Removal (CDR) · Solar Radiation Management (SRM) · Outer Space Treaty (OST) · Space law · Space policy

---

A. Koskina (✉)
National and Kapodistrian University of Athens, Athens, Greece
e-mail: akoskina@law.uoa.gr

A. Koskina · M. Plionis
National Observatory of Athens, Athens, Greece
e-mail: mplionis@noa.gr

M. Plionis
Physics Department, Aristotle University of Thessaloniki, Thessaloniki, Greece

CERIDES Centre for Risk and Decision Sciences, European University of Cyprus, Nicosia, Cyprus

K.-U. Schrogl et al. (eds.), *The Politics of Geoengineering*, Springer Climate,
https://doi.org/10.1007/978-3-032-10095-5_10

1 Importance and Relevance of the Topic

It is highly probable that in the coming decade we will irreversibly exceed the 1.5 °C global warming limit (WMO, 2023; for concerns over the effectiveness of the UNFCCC framework and its Kyoto Protocol, see Sacco et al., 2022). Therefore, there is an urgent need to further improve the understanding of climate change and to work on innovative approaches to tackle this issue. In this context, GE—broadly defined as a "process through which humans voluntarily manipulate the energy input to the Earth System deploying natural climate impacting processes by means of available technologies" (Keith, 2000; Sacco et al., 2022)—was conseptualized as a possible solution in the 60s (Sacco et al., 2022), and acquired a central role in the political agendas right from the outset (Taubenfeld, 1967). Critically, although the relevant research is still in its infancy, GE is regarded to be an option deserving to be considered, in case current mitigation or adaptation measures prove unsuccessful.

The dominant perspective (The Royal Society, 2009) is that GE may be divided—depending on its purpose—into carbon dioxide removal (CDR) techniques aimed at directly removing CO_2, and solar radiation management (SRM) to influence the solar radiation balance (also, Dicaire & Summerer, 2013). With regard to CDR techniques, emphasis until now has been placed on reducing the source or increasing the sinks of greenhouse gases on Earth; e.g., by enhancing the natural weathering processes or fertilizing oceans with naturally scarce nutrients (ibid.). Interestingly, the latest studies also established that "emissions of carbon dioxide (CO_2) caused by human activity are propagating upward to the highest regions of the atmosphere", which is likely to lead to significant adverse effects on the orbital debris environment (Naval Research Laboratory, 2012). As a result, the use of CDR techniques in/from space will most probably have to be considered in the near future, and such systems assessed as to their compatibility with the rules applicable to activities in space. On the other hand, SRM solutions include (among other things) space-based technologies, such as shields or deflectors placed in outer space to reduce the amount of solar energy reaching the Earth (Dicaire & Summerer, 2013; Schrogl & Summerer, 2016). They may additionally refer to solutions consisting in the direct alteration of the near-Earth environment (Mautner, 1991)—which should, again, comply with all the rules applying to space activities—as well as non-space-based solutions (like those aimed at increasing the Earth albedo by injecting reflecting aerosols in the upper atmosphere; Dicaire & Summerer, 2013).

Deriving from the above—namely, taking into account that some of the key features characterizing CDR techniques are the same as some of the ones characterizing SMR techniques, such as the direct alteration of the (space) environment—one may suggest an alternative approach, that is to divide space-based GE systems according to their functioning (i.e., rather than according to their purpose). Therefore, based on existing knowledge, they could be categorized into:

i. Technologies directly aimed at altering the environment's composition, such as via the direct engineered capture of CO_2 from ambient air (carbon dioxide removal) or the deployment of fine-grained dust in orbit around the Earth

(Mautner, 1991). In this case, the solution is expected from the modification of said composition; the technology being used has a secondary role as it may well consist in existing (i.e., well-known and established) space infrastructures, such as rockets for injections into higher atmospheric layers (Dicaire & Summerer, 2013; Mautner, 1991).

ii. Space-based GE solutions aimed at using new generation systems (i.e., untried and untested technology), for protecting the Earth against further deterioration: here, although an intervention on the environment's composition is attempted or foreseen, the solution is principally expected from the utilization of new, leading-edge technology, such as fleets of shields placed in low-earth orbit or at the interior Lagrange point (McInnes, 2009).

There is still a long way to go toward achieving the objective of using space-based GE. However, before putting theory into practice, a number of legal questions should be addressed. More precisely, apart from the question of which commonly agreed legal framework would ensure the viable operation of space-based GE to the benefit of all (McInnes, 2016) and in the long run (that is, once the implementation of such technologies becomes feasible), it is appropriate to first examine the conditions under which said systems could be developed and utilized, to be consistent with applicable law as well as acceptable to the international community from an ethical and legal standpoint.

2 Defining Key Legal Problems and Related Conflicts

In determining the legality of implementing space-based GE solutions, two elements should be examined: (a) GE technologies per se and (b) the possibility of using those in and/or from outer space. Interestingly, with regard to point (b), it seems that each of the two categories of space-based GE suggested above, poses different legal problems.

The primary issue—namely (a) GE technologies per se, which until now principally refers to systems created to be used on Earth—has been widely discussed. In brief, it is argued that there is no general prohibition to develop and exploit GE technologies (Proelss & Steenkamp, 2023). Indeed, the rules applying to environmental modification techniques in general—such as those adopted in the Convention on the prohibition of military or any other hostile use of environmental modification techniques (ENMOD, 1976), which is regarded as being "one of the most relevant treaties to geoengineering" (Grisé et al., 2021)—or GE technology in particular (e.g., the CBD and the London Convention and London Protocol; National Academy of Sciences, Engineering, and Medicine, 2021) do not establish any general ban on the use of GE for peaceful purposes, neither on Earth nor in outer space (GE may be, theoretically, used for military purposes; however, this is not regarded as its prime objective, see Keith, 2000). Be that as it may, legal experts also contend that the (long-term) consequences of using GE technology on a large scale are so far unknown (for albedo modification using space shields, Keith, 2000; in the same vein, Martin, 2020 and Sacco et al., 2022; for the major secondary effects of both SRM and CDR, see Grisé et al., 2021). For this reason, they tend to argue that the use of GE should not be considered to be legal, but rather unregulated (although not taking place within a legal vacuum), and that major questions of concern—including important legal challenges—arising from both their transboundary and global character should soon be addressed (Schrogl & Summerer, 2016).

That being said, the issue is further complicated in the event that GE technologies are being used in and/or from space—that is, point (b)—since all activities taking place in this context are governed by the specific rules of the Outer Space Treaty (OST, 1967) and International Space Law as a lex specialis. Hence, according to the OST and the principle of freedom of use of outer space, States would be free to use space-based GE in outer space (without discrimination of any kind) (OST, Art. I(2)); such use should be made for the benefit and in the interests of all countries (OST, Art. I(1)) on a basis of equality, in line with international law (OST, Art. I(2) and Art. III) and to maintain international peace and security (OST, Art. I(2) and Art. III). At the same time, States would bear international responsibility for GE space activities (OST, Art. VI) and remain liable for damage caused by GE technologies (i.e., in their quality of space objects) in case they launch or procure the launching of such items (OST, Art. VII). Finally, given that outer space is a res communis (Malagar & Magdoza-Malagar, 1999), States should avoid creating conditions that would prevent others from exercising their own right to freely explore and use it.

In other words, the use of GE systems in/from outer space does not seem to be a priori excluded by the OST. Nonetheless, a closer look shows that each of the two categories of space-based GE systems mentioned initially should be closely examined as to their compliance with the specific requirements laid down in Art. IX of the OST, as presented below.

Firstly, GE systems directly aimed at altering the environment's composition should be assessed in light of Art. IX, stipulating that States shall pursue studies of outer space—i.e., including the Moon and celestial bodies—and explore them "so as to avoid their harmful contamination and also adverse changes in the environment of the Earth resulting from the introduction of extraterrestrial matter" (OST, Art. IX). Despite its narrow scope of application (Hofmann & Bergamasco, 2020; von der Dunk, 2008), scholars suggest that this provision (coupled with Art. III of the OST) could be regarded as also encompassing new kinds of environmental contamination, to take into account the developments in international environmental law (Gupta & Agasti, 2022). On this basis, one could consider that there is—notwithstanding the vague wording of Art. IX (Hofmann & Bergamasco, 2020)—an undeniable demand to achieve a high level of protection for the environment as a whole (including space; Kramer, 2014). Following this reasoning, it may be argued that any utilization of space-based GE systems directly aimed at altering the environment's composition should go hand in hand with the adoption of "appropriate measures", as stipulated in Art. IX of the OST; which however does not define them (for measures proposed up until now, see inter alia Bittencourt Neto, 2013; Mustow, 2018).

Secondly, in examining the compatibility of GE solutions based on the use of untried and untested technology with the OST, account must be taken of the next rule laid down in Art. IX (i.e., when a State has reasons to believe that any space activity could cause potentially harmful interference with other States' activities, it should first "undertake appropriate international consultations"). As a matter of fact, experts argue that using this second type of space-based GE technology could well result in serious threats to the undisturbed functioning of other space objects: "the use of vast numbers ($\sim 5 \times 104$) of 100 km^2, actively controlled occulting discs in the Earth's orbit has been considered, but would likely lead to an apparent flickering of the Sun ($\sim$2 percent amplitude) and would create a significant orbital debris hazard" (McInnes, 2009). Therefore, seen from this angle, any utilization of a fundamentally new technology, especially in case of a large-scale infrastructure—e.g., large-scale GE orbiting reflectors (as a point of comparison, see the problems caused by satellite constellations in Abashidze et al., 2022)—would be likely to cause significant threats or harmful interferences with other States' activities. Extending this logic further, this kind of endeavor could arguably be conducted only following a consultation process as also laid down in said article.

Hence, analysis for both categories of space-based GE technologies shows that the OST imposes stringent rules which, however, need to be further clarified with respect to the potential use of such systems. On a more practical level, space-based GE systems have been—up to now—the subject of several debates, given the growing participation of (public and private) operators expecting a reasonable amount of legal predictability.

3 Presenting Actors and Institutions Involved Including Their Positions

Contrary to other types of GE technologies, space-based GE systems appear not to be the priority for policymakers, due to all kinds of barriers, e.g., "costs, timescales, practicalities and environmental side effects" (Proelss & Steenkamp, 2023). As a result, potential stakeholders do not seem to show much interest (critically, "only 2% of articles on GE consider space-based methods"—Baum et al., 2022), whereas the debate remains focused on "sunshades" or on "sun shields" (ibid.). In other words, space-based GE solutions are far from being discussed in a holistic way, and technologies aimed at directly altering the space environment—such as dust rings (of lunar or asteroid material) put in orbit around the Earth (Mautner, 1991)—are often left outside the scope of relevant studies, hence of political agendas.

Be that as it may, the principal actors involved in space-based GE systems are States, private stakeholders and the public; each of them operating and participating to the debate with a different focus, as they are bound by distinct priorities and constraints.

States first of all, have the sovereign right to conduct policies ensuring the protection of their legitimate national interests, and to safeguard their national wealth to the benefit of their citizens. This right is reaffirmed by the OST, in reference to space activities (OST Art. I para. 2), which must however be conducted "for the benefit and in the interests of all countries" (OST, Art. I para. 1) and "in the interest of maintaining international peace and security" (OST, Art. III). In this context, space-based GE systems raised, from the very beginning, concerns about their potential use as weapons (Kolpak, 2020)—that is, in a real or perceived military purpose (Grisé et al., 2021)—but also due to their size, as military targets (e.g., "you've got the ability to specifically target a solar shield over a particular region"—Baum et al., 2022). Above all, it was pointed out that the use of space-based GE may increase the probability of having international tensions or conflicts (Grisé et al., 2021); in this line of thinking, States gave priority to discussions on GE governance. Hence, a first step at the international level was achieved in 2019 via the proposal by Switzerland of a draft resolution for debate at the 4th United Nations Environment Assembly, entitled "Geoengineering and its governance". Said proposal asked for the preparation of an assessment of potential risks and benefits of GE technologies based on a clear CDR-SRM categorization (Switzerland, 2019), without, however, addressing more specifically any space-based GE technologies. The idea was to promote a rules-based international order, that would create an adapted governance system (Parson & Reynolds, 2021) and would allow to avoid confusion and waste of disparate and uncoordinated initiatives (Grisé et al., 2021). Overall, although no agreement was found, the initiative brought the burning issue of GE governance to the fore of the public debate (Tollefson, 2019; UNEA, 2019).

For their part, the stakeholders—who are free to conduct legal activities in the context of the market economy—remain hesitant about investing in the development and use of space-based GE systems. The reason is that research in this field was up to now mainly focused on mainstream applications of GE technology (Grisé et al., 2021), leaving the related industry with little hope that any promising space market will soon emerge. Nonetheless, despite the sharp criticism voiced by the majority of the actors involved, an optimistic view is still put forward by the very few companies working precisely on space-based methods (Baum et al., 2022).

Finally, studies show that the public (including groups and organizations) is divided between supporters and opponents of the use of GE. Among the latter, the dominant view seems to be that the development of space-based GE systems is neither practical nor desired, due to the high costs that would be spent, possibly to the detriment of existing efforts (both in climate technology and funds) aimed at tackling the causes and effects of climate change (Grisé et al., 2021). In particular, it is feared that the use of GE solutions would cancel out the efforts made by the international community in the context of the internationally developed strategy to tackle climate change, or even create serious and irreversible harm to the environment (such approaches may as well be based on religious beliefs about the environment—Baum et al., 2022). Be that as it may, the public perception on GE has still to be taken into account—e.g., fears of neocolonialism of outer space were expressed on the occasion of the adoption of the Bogota Declaration, 1976 (Gorbiel, 1978)—to allow the engagement of wider audiences in the development and use of space-based GE systems; and to avoid national governments or civil society institutions being opposed to the research or use of space-based GE, based on ethical, environmental, or risk assessment grounds (i.e., Mexico recently banned all solar GE experiments over its territory, mostly for safety reasons—see Solar Geoengineering Non-Use Agreement, 2023).

4 Policy and Regulatory Approaches and Possible Solutions

Serious concerns were early on expressed about the potential risks and on the governance and ethical implications associated with the use of GE in general, including space-based systems (the issue was raised precisely in reference to SRM technology; see IPCC, 2014). As a result, legal experts have sought to address the issue of what legal rules should apply to GE technology, giving priority to the governance challenge. The subject appeared to be a difficult one; however, in the specific case of space-based GE, a few aspects seemed, at first glance, easier to address, since such technologies would essentially have to be in line with International Space Law applying as lex specialis (i.e., given their quality of space objects and space activities).

To answer the issues raised regarding the possible models for GE governance (Reynolds, 2019)—n.b., in which the governance of GE research would not necessarily be included, since the latter raises its own special problems (on the need to rapidly address GE research governance, see C2G2, 2018; on the fact that specific attention should be given to issues pertaining to the publication of research results, see Reynolds, 2011)—experts suggested inter alia that GE activities should be carried out in accordance with the UNFCCC objectives, when possible and appropriate (e.g., States parties could incorporate GE activities in their nationally determined contributions—Reynolds, 2019). Furthermore, emphasis was put on the need to develop consensus and broad political coordination (National Academy of Sciences, Engineering, and Medicine, 2021; Zürn & Schäfer, 2013); in this respect, it was highlighted that all stakeholders (including developing countries) should be involved (Blackstock & Long, 2010), so as to reach commonly agreed sets of objectives.

Literally, all the approaches suggested above could well apply to space-based GE. Indeed, the rules and obligations found within international environmental law—or any internationally agreed rules of soft law precisely regulating GE (such as the Oxford Principles; see Rayner et al., 2013)—could apply to space-based GE on the basis of Art. III of the OST (arguably, also in conjunction with Art. IX of the OST; especially given that "Article IX does contain elements of precaution"—Proelss & Steenkamp, 2023). Nonetheless, an equally convincing approach was to propose the adoption of a specific governance framework, partly reiterating elements of existing governance schemes. Thus, one of the models suggested was that of the binding international inter-governmental agreement (IGA), regulating the International Space Station (ISS) and implemented via non-binding Memoranda of Understanding (MoU), (Schrogl & Summerer, 2016). However, a major downside of this approach would be that said agreement has only been tested in reference to a small number of countries, as it applies to only 15 States; in fact, that made "governance easier but would not necessarily meet the expectations for a climate engineering program, which should and intends to be global" (ibid.).

For the time being, in determining the applicable law to space-based GE, account must be taken of the International Space Law treaties, which establish the freedom to conduct space activity, but also regulate in substance specific legal issues. By way of illustration, the Liability Convention should apply in case of a damage being caused by a space-based GE system (i.e., when liability for space-based GE has been ascertained—Svoboda & Irvine, 2014), considering that the concept of "space object" is broadly defined therein (Proelss & Steenkamp, 2023). Yet, the principal shortcoming of this solution is that many of the adopted rules are not well adapted to the specificities of space-based GE. For instance, Art. I(a) of the Liability Convention only refers to damage to persons and property (not to damage to the environment—ibid.); additionally, no enforcement procedure was ever established in that convention (Gomez, 2012) which could be highly problematic as GE technology will probably pose challenges never seen before (e.g., damages to the ozone layer potentially caused by SRM deployment at scale—UNEP, 2023). From this perspective, applying current International Space Law rules (like those in the Liability convention) would first require agreeing upon extending and precisely adapting their scope of application (for other approaches on compensation and liability due to GE, see Reynolds, 2019).

On a more practical level, and as might have been expected, experts seem to agree that before moving further in the direction of space-based GE systems, small-scale (and at the same time, reversible and affordable) experiments should be carried out (Schrogl & Summerer, 2016). In other words, there appears to be a convergence on the view that it is necessary to secure and increase the funds for GE research (Reynolds, 2019), to gain more knowledge on all types of GE systems before going further in this direction.

5 Policy and Diplomatic Mechanisms to Reach Solutions

Experts have already underlined the need to develop a new international agreement precisely dedicated to GE (Grisé et al., 2021). In this context, the framework of International Space Law could also be revisited to consider specific aspects of space-based GE (instead of merely trying to adapt existing rules); such as to address the following key challenges.

i. *The need to debate, define and agree on the environmental goals to be achieved.*

It has been rightly pointed out that the first question is whether a venture like space-based GE—solar shields, precisely—should be organized and governed like a (typical) space project or a (typical) terrestrial one (Schrogl & Summerer, 2016). In reality, the object behind that challenge is that States must agree on the specific goals to be achieved by using space-based GE (how much dust deposited in orbit; how much solar radiation deflected etc.), so as to avoid arbitrary and harmful decisions. By comparison, all the environmental goals have been up to now considered and agreed upon on a global level, on the basis of clearly established scientific methodology and results. From this viewpoint, space-based GE essentially consists in environmental endeavors taking place in a completely new environment (i.e., outer space) unavoidably creating transboundary effects; therefore, the intermediate goals to achieve in reference to the alteration of specific space elements (e.g., quantity of solar radiation deflected) should result from the widest possible debate and consensus, and laid down in the law.

ii. *The need to adopt rules considering the specific characteristics of each GE.*

Until now, the summa divisio for GE technology is the one between CDR and SRM; however, legislative proposals and studies address GE technology without substantially differentiating between them, despite the fact that actors have expressed the view that grouping CDR together with SRM could lead to undue restrictions (Harvard's Solar Geoengineering Research Program, 2019); i.e., they have suggested that the different kinds of technology should be considered and regulated separately (ibid.). Therefore, one could argue that in line with this reasoning, a different categorization, like the one presented above in Sect. 1, would possibly be more meaningful from the perspective of space law; namely, differentiating between (i) space-based GE systems directly aimed at altering the environment's composition and (ii) those consisting in the use of untried and untested technology. The rationale behind this approach would be that the two technologies raise different concerns, since (i) would mainly create threats of environmental pollution while (ii) would essentially consist in huge pieces of satellite infrastructure in the LEO or GEO potentially posing high risks of collision (Baum et al., 2022). In other words, each category would raise challenges that have been up to now regulated by different sets of laws. Accordingly, the measures applying to each category of GE should be different and build upon the existing frameworks that are best suited to their specific regulatory needs (given that International Space Law may well operate, based on Art. III of the OST, in conjunction with other rules of international law).

The two perspectives developed above highlight that the development and use of space-based GE raise questions never addressed before. Ideally, all these issues should be discussed in detail, to allow the adoption of specific regulatory instruments (i.e., to address and regulate each space-based GE system in detail) which should, in any event, reflect the broadest possible consensus among States.

References

Abashidze, A., Chernykh, I., & Mednikova, M. (2022). Satellite constellations: International legal and technical aspects. *Acta Astronautica, 196*, 176–185. https://doi.org/10.1016/j.actaastro.2022.04.014

Baum, C. M., Low, S., & Sovacool, B. K. (2022). Between the sun and us: Expert perceptions on the innovation, policy, and deep uncertainties of space-based solar geoengineering. *Renewable and Sustainable Energy Reviews, 158*, 112179. https://doi.org/10.1016/j.rser.2022.112179
Bittencourt Neto, O. D. O. (2013). *Preserving the outer space environment: The "precautionary principle" approach to space debris.* Proceedings of the International Astronautical Congress, IAC, pp. 11213–11223.
Blackstock, J. J., & Long, J. C. S. (2010). The politics of geoengineering. *Science, 327*(5965), 527. https://doi.org/10.1126/science.1183877
C2G2 Carnegie Climate Geoengineering Governance Initiative. (2018). *Governing Solar Radiation Modification (SRM)—Policy brief*. Retrieved January 15, 2025, from https://www.c2g2.net/project/policy-brief-governing-solar-radiation-modification/
Convention on the Prohibition of Military or Any Other Hostile Use of Environmental Modification Techniques (ENMOD). (1976, December 10). *United Nations Treaty Series*, vol. 1108, p. 151. Entered into force October 5, 1978. Available at: https://treaties.un.org/Pages/ViewDetails.aspx?src=TREATY&mtdsg_no=XXVI-1&chapter=26&clang=_en
Dicaire, I., & Summerer, L. (2013). *Climate engineering: Which role for space?* Preprint submitted to 64th International Astronautical Congress, Beijing, China. Copyright by ESA. Published by the IAF. Available at: https://www.esa.int/gsp/ACT/doc/ESS/ACT-RPR-ESS-2013-IAC-ClimateEngineeringWhichRoleForSpace.pdf
Gomez, C. (2012). *The optional rules of arbitration of disputes relating to outer space activities of the Permanent Court of Arbitration, a real option for the solution of conflicts in space matter?* Proceedings of the International Astronautical Congress, IAC. Available at: https://ssrn.com/abstract=2963687
Gorbiel, A. (1978). The legal status of geostationary orbit: Some remarks. *Journal of Space Law, 6*(2), 171–178. Available at: https://airandspacelaw.olemiss.edu/pdfs/jsl-6-2.pdf
Grisé, M., Yonekura, E., Blake, J. S., DeSmet, D., Garg, A., & Lee Preston, B. (2021). *Climate control: International legal mechanisms for managing the geopolitical risks of geoengineering.* RAND Corporation. Available at: https://www.rand.org/pubs/perspectives/PEA1133-1.html
Gupta, B., & Agasti, T. (2022). The curious case of Article IX and outer space environment. *Journal of Environmental Impact and Management Policy, 2*(02), 7–25. https://doi.org/10.55529/jeimp22.7.25
Harvard's Solar Geoengineering Research Program. (2019). *Perspectives on the UNEA resolution.* Available at: https://geoengineering.environment.harvard.edu/blog/perspectives-unea-resolution
Hofmann, M., & Bergamasco, F. (2020). Space resources activities from the perspective of sustainability: Legal aspects. *Cambridge Prisms: Global Sustainability, 3*, e4. https://doi.org/10.1017/sus.2019.27
IPCC. (2014). *Climate change 2014: Synthesis report. Contribution of Working Groups I, II and III to the Fifth Assessment Report of the Intergovernmental Panel on Climate Change.* IPCC. Available at: https://www.ipcc.ch/site/assets/uploads/2018/05/SYR_AR5_FINAL_full_wcover.pdf
Keith, D. W. (2000). Geoengineering the climate: History and prospect. *Annual Review of Energy and Environment, 25*, 245–284. Available at: https://keith.seas.harvard.edu/files/tkg/files/26.keith_.2000.geoengineeringhistoryandprospect.e.pdf
Kolpak, G. (2020). *From ENMOD to geoengineering: The environment as a weapon of war.* Conflict and Environment Observatory. Available at: https://ceobs.org/from-enmod-to-geoengineering-the-environment-as-a-weapon-of-war/
Kramer, W. R. (2014). Extraterrestrial environmental impact assessments—A foreseeable prerequisite for wise decisions regarding outer space exploration, research and development. *Space Policy, 30*(4), 215–222. https://doi.org/10.1016/j.spacepol.2014.07.001
Liability Convention. (1972). *Convention on International Liability for Damage Caused by Space Objects.* Entered into force September 1, 1972. Available at: https://www.unoosa.org/oosa/en/ourwork/spacelaw/treaties/liability-convention.html

Malagar, L. B., & Magdoza-Malagar, M. A. (1999). International law of outer space and the protection of intellectual property rights. *Boston University International Law Journal, 17*(2), 311–356. Available at: https://papers.ssrn.com/sol3/papers.cfm?abstract_id=1687422

Martin, C. (2020). Atmospheric intervention? The climate change crisis and the jus ad bellum regime. *Columbia Journal of Environmental Law, 45*(S), 233–280. https://doi.org/10.7916/cjel.v45iS.5786

Mautner, M. N. (1991). A space-based solar screen against climate warming. *Journal of the British Interplanetary Society, 44*, 135–138. Available at: https://scholarscompass.vcu.edu/chem_pubs/85/

McInnes, C. R. (2009). Space-based geoengineering: Challenges and requirements. *Proceedings of the Institution of Mechanical Engineers Part C: Journal of Mechanical Engineering Science, 224*(3), 571–580. https://doi.org/10.1243/09544062JMES1439

McInnes, C. R. (2016). Near Earth asteroid resource utilisation for large in-orbit reflectors. *Space Policy, 37*, 62–64. https://doi.org/10.1016/j.spacepol.2016.07.002

Mustow, S. E. (2018). Environmental impact assessment (EIA) screening and scoping of extraterrestrial exploration and development projects. *Impact Assessment and Project Appraisal, 36*(6), 467–478. https://doi.org/10.1080/14615517.2018.1500092

National Academy of Sciences, Engineering, and Medicine. (2021). *Reflecting sunlight: Recommendations for solar geoengineering research and research governance*. The National Academies Press. https://doi.org/10.17226/25762

Naval Research Laboratory. (2012, November 11). *Scientists detect carbon dioxide accumulation at the edge of space. Phys.org*. Retrieved January 12, 2025, from https://phys.org/news/2012-11-atmospheric-co2-space-junk.html

Outer Space Treaty. (1967). *Treaty on Principles Governing the Activities of States in the Exploration and Use of Outer Space, including the moon and other celestial bodies*. Entered into force October 10, 1967. Available at: https://www.unoosa.org/oosa/en/ourwork/spacelaw/treaties/outerspacetreaty.html

Parson, E. A., & Reynolds, J. L. (2021). Solar geoengineering: Scenarios of future governance challenges. *Futures, 133*, 102806. https://doi.org/10.1016/j.futures.2021.102806

Proelss, A., & Steenkamp, R. C. (2023). Geoengineering: Methods, associated risks and international liability. In P. Gailhofer, D. Krebs, A. Proelss, K. Schmalenbach, & R. Verheyen (Eds.), *Corporate liability for transboundary environmental harm* (pp. 207–236). Springer. https://doi.org/10.1007/978-3-031-13264-3_9

Rayner, S., Heyward, C., Kruger, T., Pidgeon, N., Redgwell, C., & Savulescu, J. (2013). The Oxford principles. *Climatic Change, 121*, 499–512. https://doi.org/10.1007/s10584-012-0675-2

Reynolds, J. L. (2011). The regulation of climate engineering. *Law, Innovation and Technology, 3*(1), 113–136. https://doi.org/10.5172/lit.2011.3.1.113

Reynolds, J. L. (2019). Solar geoengineering to reduce climate change: A review of governance proposals. *Proceedings of the Royal Society A, 475*, 20190255. https://doi.org/10.1098/rspa.2019.0255

Sacco, N., Janssens-Maenhout, G., Galmarini, S., & Michel, Q. (2022). *Geo-engineering: A roadmap towards international guidelines* (JRC Technical Report). European Commission. Available at: https://publications.jrc.ec.europa.eu/repository/handle/JRC99777

Schrogl, K.-U., & Summerer, L. (2016). Climate engineering and space. *Acta Astronautica, 129*, 121–129. https://doi.org/10.1016/j.actaastro.2016.06.036

Solar Geoengineering Non-Use Agreement. (2023, January 20). *Experimentation with solar geoengineering will not be allowed in Mexico*. Retrieved January 15, 2025, from https://www.solargeoeng.org/experimentation-with-solar-geoengineering-will-not-be-allowed-in-mexico/

Svoboda, T., & Irvine, P. J. (2014). Ethical and technical challenges in compensating for harm due to solar radiation management geoengineering. *Ethics, Policy and Environment, 17*(2), 157–174. https://doi.org/10.1080/21550085.2014.927962

Switzerland. (2019, February 19). *Resolution for consideration at the 4th United Nations Environment Assembly—Revised version. Geoengineering and its governance.* Available at: https://geoengineering.environment.harvard.edu/files/sgrp/files/draft_unea_resolution.pdf

Taubenfeld, H. J. (1967). Weather modification and control: Some international legal implications. *California Law Review, 55*(2), 493–506. https://doi.org/10.2307/3479357

The Royal Society. (2009). *Geoengineering the climate: Science, governance and uncertainty*. The Royal Society. Available at: https://royalsociety.org/-/media/policy/publications/2009/8693.pdf

Tollefson, J. (2019). Geoengineering debate shifts to UN environment assembly. *Nature, 567*(7747), 156. https://doi.org/10.1038/d41586-019-00762-7

United Nations Environment Assembly (UNEA-4). (2019). *Summary report*. Available at: https://enb.iisd.org/events/4th-meeting-oecpr-and-4th-session-unea/summary-report-11-15-march-2019

UNEP. (2023). *One atmosphere: An independent expert review on solar radiation modification research and deployment*. Available at: https://www.unep.org/resources/report/Solar-Radiation-Modification-research-deployment

von der Dunk, F. (2008). *Legal aspects of using space-derived geospatial information for emergency response, with particular reference to the Charter on Space and Major Disasters*. Space and Telecommunications Law Program Faculty Publications, Paper 14. Available at: http://digitalcommons.unl.edu/spacelaw/14

World Meteorological Organization. (2023). *WMO global annual to decadal climate update target years: 2023 and 2023–2027*. Available at: https://library.wmo.int/idurl/4/66224

Zürn, M., & Schäfer, S. (2013). The paradox of climate engineering. *Global Policy, 4*(3), 266–277. https://doi.org/10.1111/gpol.12004

Societal and Public Aspects

Geoengineering and Public Value: Toward a Global Climate Sensibility

Rick Wylie

Abstract Currently characterized by complexity, uncertainty, and controversy the future of potentially vital yet risky geoengineering (GE) technologies is far from certain as even at-scale research is presently lacking. Moreover, there is no governance framework at an appropriate scale to provide oversight and control over GE. Given the complexity of G an anticipatory form of governance will be an essential precursor to the investigation and implementation of GE. Such a governance architecture requires agility given the uncertain impact and interaction between GE initiatives upon global climate systems which create new relationships between allies and adversaries. A governance framework based upon the globally held (though individually prioritized) human value set could accommodate a diversity of interests and a human perspective on the impact of GE on the common good at a range of scales. A public value perspective could, it is suggested, create the global sensibility necessary for the emergence of a global public sphere grounded in the everyday lives of individual citizens across the globe.

Keywords Public value · Public sphere · Globalization · Governance

1 Introduction

In this chapter we consider the concept of public value as a framework for exploring issues relating to GE policy and governance. It is suggested that insights from this analysis could inform the development of a future research program associated with the implementation of GE. The chapter begins by unpacking the concept of public value and explores what it can offer as a perspective on the GE policy debate with a focus upon the concept of human values and the public sphere. The chapter considers some of the key issues associated with the implementation of GE (and research at scale) especially (though not limited to) SRM, discussing the insights a public value approach gives to its governance. It concludes by drawing upon Ostrom's Institutional

R. Wylie (✉)
University of Lancashire, Preston, UK
e-mail: rwylie@lancashire.ac.uk

K.-U. Schrogl et al. (eds.), *The Politics of Geoengineering*, Springer Climate,
https://doi.org/10.1007/978-3-032-10095-5_11

Analysis and Development (IAD) in a framework encompassing local, national, and global levels to address some of the key issuesand relationships which GE poses to the global society and which links individual citizens with global issues in a polycentric policy environment and which may, it is argued, create a global narrative necessary to address the problems, potential, and practicalities of GE.

In this discussion on GE, we can't use 'final language' as it is much too early to be definite about either deployment or impact, or the actors involved in it. Currently, GE, especially solar geoengineering (SGE), is at the earliest stage of a lengthy pre-implementation and research process, and the impact of a substantive at-scale deployment is unknown. Indeed, as argued by Rayner et al., '*...at present, all geoengineering technologies are speculative and extensive research into their technical, environmental, socio-political, ethical and economic characteristics is necessary before their use could be sensibly contemplated...*' (Rayner et al., 2013, p. 501). This uncertainty is compounded by complexity associated with unknowns in changes in the Earth's weather systems developing as the effects of human-induced climate change are becoming more intense. Moreover, GE is a very controversial topic and even research into SGE at a small scale has been challenging and initiatives challenged, and carbon reduction management (CRM) has massive cost implications for implementation at any relevant scale which may have significant tax or market-based resourcing consequences.

Presently there is little coherent governance of GE at a scale embracing local interests and global processes. This discussion will focus on a framework for an anticipatory, agile mode of governance which can be used flexibly as GE develops politically, scientifically, and publicly. Indeed, '*...there is currently no dedicated, comprehensive international framework or fora to inform, guide and connect governance processes for SRM research, development, demonstration, or deployment. This poses serious risks given that SRM would affect every country on the planet, but not necessarily equally.*' (Carnegie Institute for Climate Change 2022).

Today, in respect of GE '*...calls for a governance framework abound...*' (Dilling & Hauser, 2013). The subject of this chapter is therefore how the concept of public value governance can inform the development of an appropriate governance framework for GE.

Before discussing the problems and solutions, the concept of public value is introduced, itself a 'portmanteau' topic with many interpretations (for a more comprehensive introduction see Wylie et al., 2020). Essentially, public value means the realization of value in the public sphere as perceived by individual citizens seen through the lens of their values, their needs, and their motivations (Wylie, 2021). Public value is a relational and dialogic concept; it is about how we interact, contribute, and draw from the world we live in. Public value exists in the public sphere, and in the spaces we live our daily lives. It's about the prosaic, the commonplace, the everyday. Conceptually, public value connects the micro level of the individual citizen with the macro level of society and the public sphere. As will be shown, it is a human-centric concept about individuals' relationships with the wider world and each other.

In his seminal book *Creating Public Value* (1997), Mark Moore drew attention to the importance of the concept of public value theorizing and operationalizing how

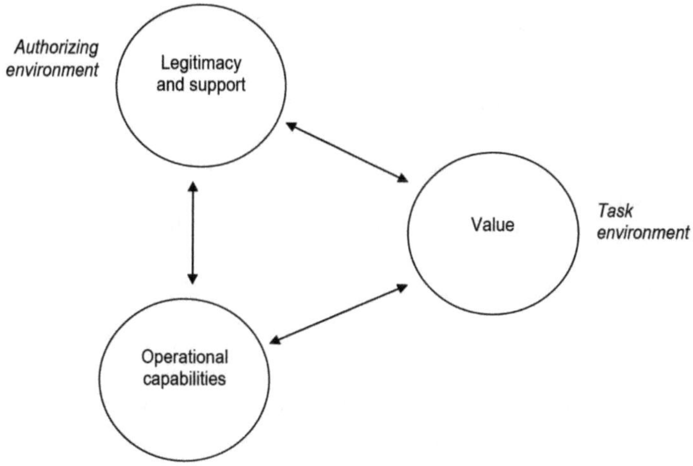

Fig. 1 Moore's Triptych, the public value strategic management triangle (After Moore, 1997)

organizations can contribute to society and to the common good. Moore's approach to the strategic management of public value is usually rendered in the form of following figure which forms a useful heuristic in discussing public value in the context of policy and governance.

Figure 1 shows the three key elements of strategic public value management and their fundamental relationship. In essence, organizations (in all sectors) secure legitimacy to win tax resources (or public support) to build capacity (alone or severally) to achieve public value and contribute to society. Fundamentally, that iswhat public value is from a policy management standpoint; activities to contribute to the common good by providing society with what individuals value. From this public value management standpoint an effective strategy must, firstly aim at achieving something that must contribute value; secondly be legitimate and politically supported in the authorizing environment; and thirdly be do-able within the capacity of the organization in the context of its operational setting and maybe within the context of partnership or collaborative arrangements. Moore's strategic triangle connects the micro level of the individual citizens with the context of polycentric policy communities and more widely to the public sphere.

2 Defining Key Problems

GE from a public value perspective has several overarching problems including complexity, scale, and controversy esp. Solar Radiation Management (SRM) and these problems are compounded by uncertainty, widespread concern and the dynamic nature of GE development and future impacts. Indeed, GE, especially SRM has many of the characteristics of a so-called wicked problem, following Geuijen et al. (2017)

it evokes several issues all of which have implications for the effective governance of GE:

- Multiple, potentially conflicting values;
- Strong political passions on all sides of the issue;
- Substantive uncertainty on how best to solve the problem; and it creates;
- multiple arenas for social deliberation and action, a polycentric policy environment with multiple centers of authority and it may be that multiple tiers of governance will be required to address and respond to the issues associated with the implementation and impact of GE (Paschke et al., 2019).

A particular problem with the governance of SRM is the impact on natural weather systems and consequently on societies and the potentially patchy nature of its deployment and subsequent impact. For CDR, a key issue could be the financial cost for at-scale implementation to secure the global impact it generates with a perhaps significant financial burden on contributor states and societies.

Another topic is the polycentric nature of GE as a policy issue and the need for multi-level, multi-sector (probably) and multi-scale solutions. GE requires local and global perspectives and policy to accommodate the ability to respond and react both locally and nationally to the impact of deploying GE technologies on the climate and consequential. Moreover, the actors and institutions associated with GE deployment must be sufficiently agile to respond to 'knock-on' impacts in other localities (good or bad) on a global scale.

From a governance standpoint a framework should, it is argued, be in place before deployment of GE (Rayner et al., 2013) and yet the actual scope of the governance requirement is uncertain. Though what appears to be certain is that GE will become a dynamic issue, which will require a governance architecture that is both anticipatory and agile. It is already a controversial issue for SRM—even small-scale research effects (like SCOPEX, for example Oksanen, 2023) and the wider impacts of SRM are currently known only from limited modelling. For example, in the context of SRM, there have been no-large scale experiments. Moreover, the deployment at scale, given its significant potential impact, further adds to the uncertainty surrounding its implementation.

It is probable that the deployment of GE—especially SRM—will see both winners and losers. Moreover, it is expected that the issue will create geopolitical impacts on the global level, creating both new conflicts and collaborations between states and societies. It is also likely that new actors will bring relatively low cost SRM technologies—easily affordable by an IT billionaire, for instance! Research on the stratospheric injects for SRM, for example, is especially problematic, given its unknown spatial impacts and long-term consequences.

As Greely (2020) remarked

> The climate system is global and immensely complex, and changes to the climate system affect people, businesses, and governments around the world. Currently, he said, there are no international entities, including professional organizations, that can credibly issue guidelines for SG research that will be respected globally. There is also an absence of international agreement on the governance of SG research.

Scale of impacts is an important issue in respect to the GE implementations. As is argued by Pidgeon et al.,'*...aerosols (are) also seen to raise significant problems of international governance and control, underlining the importance of current efforts to develop research governance structures*' (2013, p. 459). While considering the uniqueness of SG research, Jasanoff (2020, p. 2) observed that climate change challenges the scales of human organization and self-understanding on multiple dimensions relevant to governance (e.g., space, time, community, and polity). She notes that this has implications for questions about 'who governs' and by what means. From her perspective, the framing of research questions is itself value-laden and a variety of judgments determine which questions are or are not included in the research program. Jasanoff further emphasizes the importance of language and argues thatthe idea of a public is a construct—publics 'are made, not found'—consequently, ideas about proof and demonstration are culturally conditioned.

Finally, the polycentric nature of the operational environment poses questions about the effective governance of GE given the potentially multiple, competing, centers of operational capacity and the multi-scale, multi-sector nature of GE policy and implementation. The policy, technology, social geography, and geometry of the issue are potentially extremely complex. Indeed, it is widely acknowledged that the challenges of GE are as much political and social as they are technical (Rayner et al., 2013).

Clearly, a governance architecture for GE must be sufficiently agile and have the potential scope to accommodate national interests, regional and local communities, and a polycentric mix of sectors and actors—across multiple levels. The evolving nature of the issue also requires a governance architecture to accommodate a plurality of interests and a diversity of outcomes and engage multiple value standpoints given the uncertain, patchy and dynamic nature of GE impacts and outcomes across the globe.

3 Highlighting the Importance of the Topic

Given the issues outlined above, a public value approach can offer important insights into a governance framework for the implementation of GE. As discussed, GE requires a governance framework that is both agile (able to cope with chance) and anticipatory (credible and robust before implementation). Given the evolving nature of the impacts of implementing this uncertain, unproven technology and given the possible requirement for a governance framework to be in place before the research and roll-out of GE technologies is permitted. Given this anticipatory dimension, any proposed governance framework needs to be inclusive to accommodate a plurality and diversity of interests interacting with an uncertain and even dangerous technology.

In this context a public value framework has something to offer GE governance and to appreciate this it is necessary to unpack and extend the application of Moore's framework to the GE debate to accommodate spatial variance and value diversity.

To achieve this, works of two scholars that contributed significantly to the concept of public value and its application are utilized. John Benington's work on the public sphere (Benington, 2011) and Timo Meynhardt's psychological approach to public value with its essential foundation of human needs and values (Meynhardt, 2009). Organizations only create value for society if they contribute effectively to the common good. As Meynhardt argued (2009) '*...public value represents what individuals' value against the background of their own and the public's interests.*' The leading work of Schwartz (2012) serves to appreciate the full gamut of human values providing a comprehensive conceptual framework necessary to appreciate, articulate and accommodate the implications of value diversity in GE implementations.

There is also an issue in identifying and accommodating 'the public' in respect of both the costs, benefits, and control of GE implementation. Public value is value derived from and for the public, but who are the public? Public value is about adding value to the public sphere, and therefore a public value perspective upon GE must centrally address the concept of 'the public'—the beneficiaries of value.

Following Dewey (1927), Moore suggests the importance of 'calling a public into existence.' From a public value standpoint, a 'public' is created rather than found, and for each public good and policy issue there corresponds some 'public.' Given the spread of global IT and digital networks this is especially the case for GE which could create complex implementations and impacts encompassing a constant flux of global discourses, regional implementations, and local impact and behavior. John Dewey has formulated the concept 'the public consists of all those who are affected by the indirect consequences of transactions to such an extent that it is deemed necessary to have those consequences systematically provided for' (Ostrom et al., 1961) citing John Dewey, *The Public and Its Problems* (New York, 1927).

Operationally, however, identifying who and what the 'public' is can be problematic (Frederickson, 1991). Nonetheless, Meynhardt (2009) sees the 'public' is an '*indispensable operational fiction necessary for action and orientation in a complex environment.*' In other words, as complexity increases the more '*the public*' becomes a social construct '*necessary for acting, but hard to pin down*' (cited in Bryson et al., 2014, p. 204). This inherent flexibility of the concept of the public gives a public value framework-based governance approach useful purchase on the issue of GE.

An expanded strategic triangle set in the context of the public sphere accommodating the expanded concept of values is rendered in Fig. 2.

Following Moore's strategic triangle as a 'heuristic' to inform the investigation is the idea of human values to understand the process by which individuals interpret and understand the world they live in (their public sphere, as it were). Human values may be seen as psychological drivers that determine the orientations and behavior evoked in response to an individual's interactions with the world they inhabit—they are guiding principles in an individual's life (Schwartz, 2012) which they use to evaluate things.

But since the impact of GE may be patchy it may be that the value implications of GE are also variable and that to accept a wider implementation of GE (esp. SRM) at an appropriate scale various impacts on individuals—some positive, others negative—must be accommodated. These impacts will affect or resonate (Rosa, 2019)

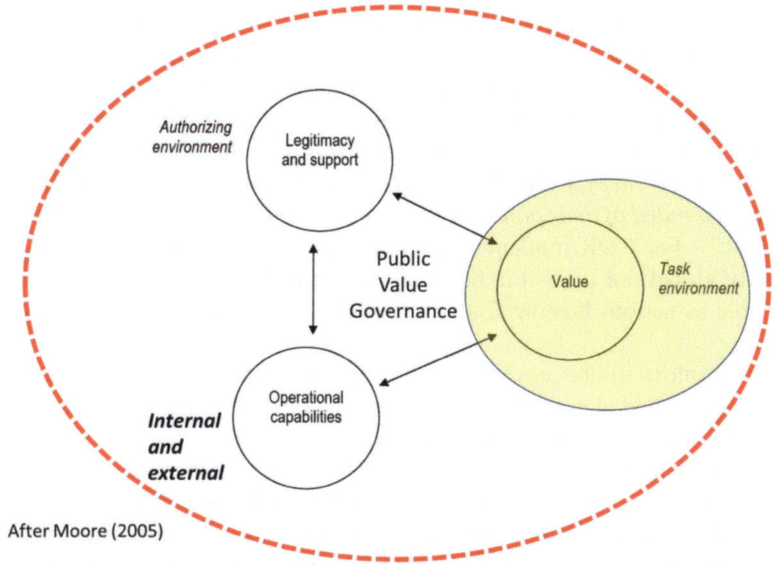

Fig. 2 The expanded strategic triangle rendering the Moore's triptych in the context of a public sphere (After Moore, Benington and Meynhardt)

differentially with their values. Schwartz (2012, see also Schwartz et al., 2012) identifies so-called four 'higher order' categories of human value categories within which are values primarily concerned with a *self or personal focus* and those concerned (and still salient to the self) with a wider *societal or social focus*. In more detail these value sets are:

- **Social Focus**
  - Values that emphasize concern for the welfare and interests of others—universalism and benevolence; and
  - Values that emphasize order, safety, self-denial, preservation of the past and resistance to change—tradition, conformity, security.
- **Personal Focus**
  - Values that emphasize pursuit of one's own interests and relative success and dominance over others power, achievement, resources, dominance; and
  - Values that emphasize independence of thought, action and feelings and readiness for change—self direction in respect of thoughts and actions, stimulation, hedonism, and enjoyment.

Thinking about so-called 'wicked' problems like Human-induced climate change (HICC) the public discussion about an issue may oscillate from one value 'set' to another. In the event that SRM is used, one location may benefit from the intervention, becoming a 'winner,' while another region may become a 'loser,' experiencing worse climatic conditions as a result of the moderated climate in an adjacent region. Clearly, winner and loser issue publics may emerge with one narrative, emphasizing one value set over another dominating; however *'...the other never completely disappears; it may reemerge in times of crisis or other contextual shifts. This happens because both contain central values that are important to most people: human rights as well as national interests' matter. Hardly anybody discards one of those completely, at least not when reminded of their potential importance...'* (Geuijen, 2004, cited in Geuijen et al., 2017). For CDR implementation at scale—the climate impact of which is inherently global not local- the financial burden of this is potentially massive and inequitable as nations hosting CDR facilities bear a tax and cost burden for global benefit.

In the context of the governance of climate change and GE, a public value approach, grounded in human values provides a common frame of reference, though the priorities (and the expression of these value categories), may be different across settings and instances. However, they are universally recognized and respected and provide a universal language (Schwartz, 2012) which for the resolution of HICC through GE informed by a public value approach may be significant, allowing for the development of a values-based narrative within a governance framework.

The importance of a public value perspective on the politics of GE is beginning to emerge from this discussion. Moore's strategic triangle has been located within the context of the concepts of the public sphere and human values. The concept of the public sphere gives public value flexibility in application for the emerging issues associated with GE implementations, while the concept of human values—and the universal human values set—facilitates a wider, flexible view and vision of GE implementations by focusing upon values of *all* affected interests and individuals living in communities as emerging publics at various scales and with different impacts and perspectives on HICC and GE.

4 Actors and Positions—Toward Regulatory and Policy Approaches

GE is a complex and barely begun technical intervention with uncertain (though perhaps profound) positive and negative implications for publics scoped at local, national, and global levels. This complexity is compounded by the lack of research into the impact and effect of GE, the uncertain, evolving nature of the climate problem and the interrelationship between anthropogenic GE interventions and natural processes which are themselves evolving as humanity's carbonization of the earth's atmosphere continues. Moreover, in implementations of GE it may be that differential impacts will create 'winner' and 'loser' publics who may emerge as GE technologies are implemented.

The public value perspective on the topic of GE developed in this chapter suggests that an acceptable and agile framework needs to be in place *before* any implementation, and even research at scale, takes place. Gupta et al. argue that the governance of solar GE as *an '…anticipatory challenge here because the very contours of the 'object of governance' remain uncertain and largely even unknowable'* (2020). Clearly, the development of an acceptable governance framework relies upon wider public engagement and, in the era of so-called 'post normal' science there is clearly a requirement for public actors (and their representatives) to be brought into what was once the exclusive domain of scientists (Funtowicz & Ravetz, 1993).

Currently, at the crucial global level… *'There is currently no dedicated, formal international framework or fora to guide and connect governance processes for SRM research, development, demonstration, or deployment, which poses risks, given, for example, that SRM would affect every country on the planet, but not necessarily equally.'* (Carnegie Climate Governance Initiative, 2022).

Given that in the HICC and GE debates there is no guardian of 'public value' for this global issue, a global public sensibility embracing a diversity of value positions may emerge through exposure and contribution to a HICC discourse leveraged by (potential) solutions like GE. Global initiatives are authorized and resourced by nation states and yet there is no coincidence between a global public sphere and the global authorizing environment. In short, who acts on behalf of the global public? Moreover, the concept of GE will clearly engage with issues in the geopolitical sphere and it may be that:

> …discourse may intersperse technical and scientific arguments with more generally accessible narratives that fit together the specialists' arguments with accounts of events, emblematic cases, and even doomsday scenarios to generate compelling stories about the causes of current problems, what needs to be done to remedy them, and how they fit with the underlying values of the society… (Schmidt, 2008, p. 309)

In this emerging agenda the direction of influence may not flow from political elites but from the discursive interactions of social activists in national and local arenas leveraged by the global media and NGOs (see Keck and Sikkink, 1998). Through this discursive process a global sensibility which accommodates a diversity of value positions may develop as *'…(t)he interactive processes of discourse … exert a causal influence beyond what discourse does in representing ideas. Most generally, discourse serves not just to express one set of actors' strategic interests or normative values but also to persuade others of the necessity and/or appropriateness of a given course of action…'* (Schmidt, 2008).

In short, how do all affected interests and 'publics' identify the salient dimensions of public value in the context of decision making with global implications (and local and national implications, too)? In other words, how do public values (the adoption of human value categories within organizations strategies as end goals) emerge in the decision making processes and policy decisions of sub-global policy actors? From the local level where the impacts of HICC and GE are and will be experienced directly and where public value is realized and appreciated directly by citizens, through national level where state level actors create the authorizing environment within which local decisions are authorized, to a global public sphere in which '*...non-governmental actors become the advocates of the needs, interests and values of people at large...*' (Castells, 2008).

A public value informed approach to a governance architecture allows us to move beyond focusing on local and national processes of creating public value, to include the decisions of business, civil society, and governments at international and global level reflecting the polycentric reality of climate issues. Thus, we can get a proper, public perspective on GE embracing those who enjoy the benefits of GE implementation and those who suffers the costs for the (wider) common good. But whose needs count? Who speaks for whom? Whose voice is heard? And who co-decides at all levels, sectors, issues, and moments in time when public value gets created for some and destroyed for others? This raises the question, how can we change governance arrangements to include all affected interests?

This would amount to stimulating a global public sphere, with local as well as global platforms in which problems would get nominated for public attention and collective action, and the best means for dealing with the problems debated and enacted. It is at this level that the important question is how and who constitutes the relevant public and acts as the guardian for global public value, creating a forum where the issue is discussed, and actions agreed upon. In the context of human-induced climate change and GE—which cannotreally be separated—this becomes critically important (Geuijen et al., 2017).

Fundamentally, ecological and biophysical processes and systems do not coincide with human policy arenas and communities. As noted by Castells, there is, as with many issues in a globalized world '*...a growing gap between the spaces issues arise in and the spaces where issues are managed*' (2008). This disconnect highlights the need for innovation to address institutional voids at the transnational level. As Geuijen et al. suggest, '*When* public value *is applied to global wicked issues it offers an opportunity to explore which kind of institutional innovation is required to convene an appropriate authorizing structure in the institutional void at the transnational level*' (2017).

A public value approach to the governance of GE reveals the importance of the *authorizing environment* as '*government has a special role to play as a guarantor of public values, but citizens as well as businesses and non-profit organizations also are important as active public problem solvers...*' (Bryson et al., 2014). But from an institutional standpoint a key issue with GE is the absence of actors able to form an 'authorizing environment' at the global level. It may be argued, however, that given the broad scale of GE, a global discourse based upon the common perspective of public value may create innovative collaborations across sectors, countries, and communities which, with an appropriate institutional framework, could accommodate diverse value perspectives. This possibility is explored in the next section, where a framework for a research program is proposed.

5 Policy and Diplomatic Mechanisms—Toward a Research Program

In the overview of public value governance in the context of GE we have drawn heavily upon Moore's public value strategic management model with its focus upon three essential elements of public value strategy—capacity, legitimacy (and resources) and ultimately value. Several issues have emerged from this discussion which suggest the importance of research into this topic as an integral part of a research program for public value and GE. With its potential global impact creating both benefits and burdens upon regions (financial for CDR at scale and climate impacts for SRM with regional and wider scale implementations) a governance (and research) agenda must accommodate a diversity of perspectives and levels as GE creates winners and losers through the patchy nature of its implementation and through the various stages of its implementation.

Perhaps the principal issue in the governance of GE is its development in the absence of a global public sphere. GE decisions at all levels create value—and potentially cause value loss—across the global public sphere. In a polycentric, multi nodal, multi-sector, multi-functional world local 'context specificity' will surely be a key factor in the acceptability and acceptance of the development of policy positions to create multi-level and sectoral solutions (Geuijen, 2004) at an appropriate scale to address HICC. In this context, a GE solution to HICC—if it is to be explored, developed, and implemented—needs to accommodate global and national levels yet be grounded in localities, in the sphere of individual experience.

Previously we have discussed some of the key issues confronting the implementation of GE seen from the perspective of public value. Any policy and diplomatic mechanism must be dynamic in that the impact and implications of GE will develop as it is implemented, and possibly as climate consequences and community conflicts emerge. Therefore, a policy mechanism must be agile. The governance architecture we are proposing here as a basis for a policy research program is a flexible framework which accommodates the future evolution of GE. It also accommodates the key elements illuminated by a public value approach with its focus on the common human values set and the relationship between lived experience in the public sphere (at various levels) and policy at all relevant levels in a dynamic environment.

Drawing on previous work (Wylie, 2021) we introduce Ostrom's IAD framework (Crawford & Ostrom, 1995; Ostrom, 2011, 2019) for an insight into the application of public value research to a framework for the development of GE governance accommodating multiple levels of policy design and implementation necessary to implement a GE solution (or ultimately not implement GE as the case may be). To this end we draw upon Ostrom's Institutional Analysis and Development (IAD) framework described as '…one of the most distinguished and tested frameworks in the field of policy sciences…' (Carlsson, 2000). IAD is essentially a *framework*, a metatheoretical conceptual map which *'identifies, categorizes, and organizes those factors deemed most relevant to understanding some phenomenon'* (McGinnis, 2011).

The IAD framework comprises a general set of variables that can be used to analyze all types of institutional arrangements (Ostrom, 2011), it is especially relevant to this discussion of a public value perspective on GE in that it allows us to locate key elements associated with public value in the implementation and evaluation of policy. It offers many possibilities for a research program which is essential given the need for flexibility and agility in the governance of GE (Geuijen et al., 2017). IAD in a multi-level form creates an institutional setting for dialogue between actors, interests and levels which can accommodate multiple value perspectives, connecting actors with a range of possibly shifting positions and perspectives emerging as the technology, its impact, and the relationships between communities, nations, and technologies impacts on the climate emerge.

A key feature of the IAD Framework relates to its ability to analyze behavior at multiple levels of analysis by shifting from one action situation 'level' to deeper rule-changing situations (Kiser & Ostrom, 1982). Three nested levels are commonly recognized: the operational, collective choice, and constitutional choice (Whaley and Weatherhead, 2014). From a public value standpoint, the IAD framework is also relevant to this discussion of GE as it allows us to accommodate the concept of human values and ultimate public value at several points, including the characteristics of communities, the interests of actors, and key evaluative criteria.

As discussed, public value draws attention to the lived experience of citizens in situ and using the IAD framework is especially relevant to this discussion of GE and public value as it can be configured as a series of levels of analysis. Thus, we can accommodate the micro level of individual citizens and the lived experience of the public sphere in the context of multiple levels (e.g., national) which sets the institutional context of interactions and authorizes actions for policy actors at sub-national levels. IAD also accommodates the contextual conditions (including biophysical, social, economic, and cultural) which shape institutional design and which influence outcomes. IAD contains no normative presumption in respect of actual institutional arrangements giving researchers and analysis a comprehensive set of rules in use which shape and constrain the activities of actors in the action situation. Again, this level can accommodate human values; however, research into the connection between IAD and public value is very limited.

The IAD framework is particularly well-suited for GE implementations and policies, as it is sensitive to the needs of local conditions. The framework considers the wider world—the impact of external variables on the action arena. At any given level of analysis, it includes consideration of the ecosystem considered (biophysical conditions) and the characteristics of the community which manages the ecosystem (Ostrom, 2005). A key feature of IAD is the fact that the decisions of a given level basically set the institutional rules in use of the next, lower level. A constitution setting the basic institutional rules for a legislature, while a legislative statute sets the basic rules governing agency permit decisions. Consequently, it is primarily the decisions of the local operational level that directly affect citizens; the decisions of higher levels are guidance to lower levels. An IAD framework tailored to this discussion of public value and GE is rendered below.

The figure reveals some of the key issues we have discussed in a public value approach to the governance of GE. At any given level it relates antecedent conditions including the biophysical environment (on the left side at each level) which inform actors and their interactions which create outcomes clearly, for GE this is an essential aspect of a dynamic situation. The IAD framework is sufficiently flexible to accommodate the emergence of the issue through multiple levels and settings. Key to this discussion is Ostrom's concept of 'evaluative criteria' which address both the action arena and the outcomes: this may be key to a public value implementation of this model.

A public value approach to the politics and policy of GE drawing on the IAD framework illuminates the necessity of multiple level interactions within the global political-economic context associated with the governance of GE. Figure 3 reveals the institutional void at a global level (represented as a query) in place of 'Rules in use' at the global level. As Geuijen et al. argued, *'It is hard to envision an authorizing environment at the global level that – like government at the local and national level – can and will facilitate all affected interests having a more or less equal opportunity to voice their needs through experimentation and deliberation. There is no global authorizing environment which governs money and authority to address those needs in an equal way in any foreseeable future'* (2017, pp. 635–636).

As a framework for a public value approach to the governance of GE, IAD gives important insights into a research program based upon an agile, anticipatory architecture engaging multiple levels of localities in a global public sphere. Within this framework, the concept of public value—with its foundation in the global human value set -could provide an integrating dynamic toward a global perspective (facilitated by *inter alia* internet, media, and transnational commerce). A wider public value narrative could lead to the appreciation and ultimately acceptance of diverse value positions accommodating the wide range potentially of value positions in respect of GE and reflecting the dynamic nature of global civil society. This public value narrative may lead to the evolution of a global climate sensibility, engaging local communities with distant strangers, meshing local and wider notions of the public sphere, and reflecting the interconnected reality of global weather systems necessary for the design and implementation of GE to address the ultimately anthropogenic issue of human-induced climate change.

From the standpoint of a research programmer into the politics of GE it is evident how a public value informed approach to institutional design can suggest institutional innovations such as discourse coalitions, centered around the fundamental commonalities of the human value set, to begin to shape an 'authorizing environment' within the institutional void at the global level which bears upon the implementation of GE. From an anticipatory governance perspective, the IAD framework, expanded to include geopolitics and discourses at multiple levels may assist researchers in theorizing and articulating links between the everyday experience of individual citizens, and the discourse necessary to create an agile, governance framework, fostering institutional innovation to accommodate a dynamic issue in a polycentric, global policy environment.

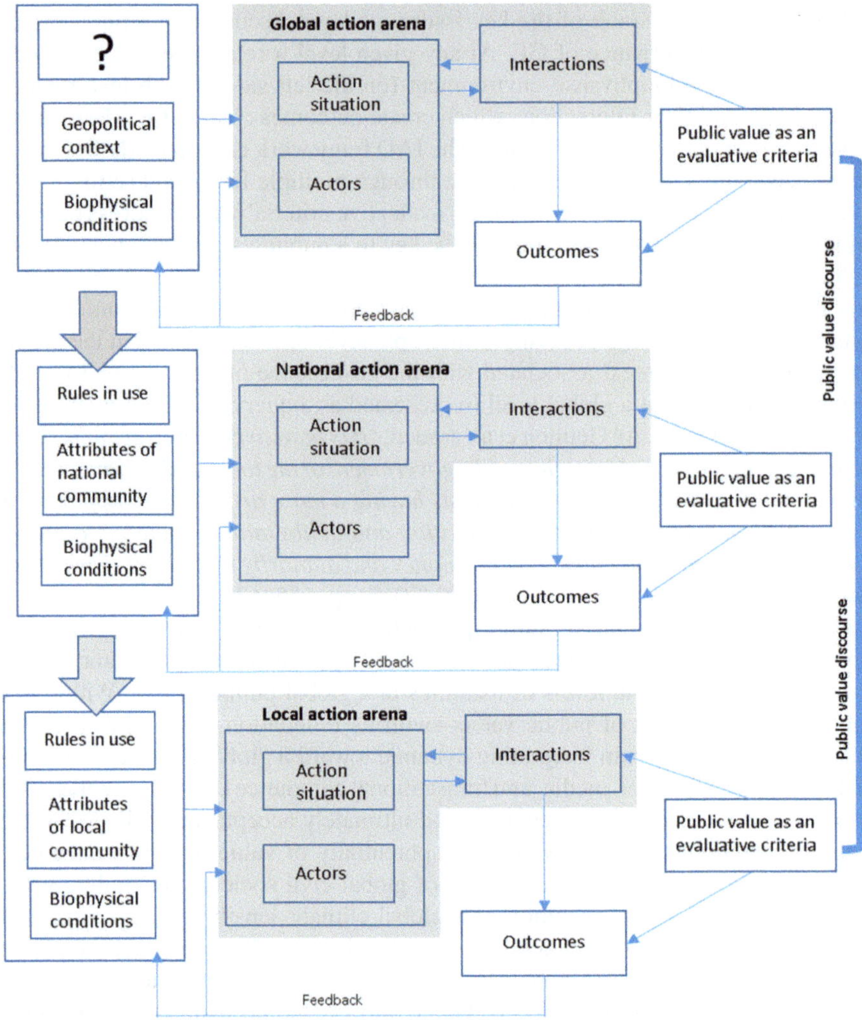

Fig. 3 A Public value informed rendering of Ostrom's IAD framework at three levels of analysis suggesting the lack of 'Rules in use' at the global level

It may be that the implementation of a public value concept in the context of global governance may foster the development of a global sensibility, perhaps a small step toward the ultimate realization of a broad commonwealth of human dignity (Lasswell, 1986) in the global public sphere.

References

Benington, J. (2011). *From private choice to public value* (pp. 31–51). Public Value.
Bryson, J. M., Crosby, B. C., & Bloomberg, L. (2014). Public value governance: Moving beyond traditional public administration and the new public management. *Public Administration Review, 74*(4), 445–456.
Carlsson, L. (2000). Policy networks as collective action. *Policy Studies Journal, 28*(3), 502–520.
Carnegie Climate Governance Initiative. (2022, August). *Briefing note on solar radiation modification what does the Intergovernmental Panel on Climate Change (IPCC) tell us?* https://www.c2g2.net/wp-content/uploads/C2GBrief_SRM_IPCCAR6-2.pdf
Castells, M. (2008). The new public sphere: Global civil society, communication networks, and global governance. *The Annals of the American Academy of Political and Social Science, 616*(1), 78–93.
Crawford, S. E., & Ostrom, E. (1995). A grammar of institutions. *American Political Science Review, 89*(3), 582–600.
Dewey, J. (1927). *The public and its problems*. Swallow Press (1954, p. 219).
Dilling, L., & Hauser, R. (2013). Governing geoengineering research: Why, when and how? *Climatic Change, 121*, 553–565.
Frederickson, H. G. (1991). Toward a theory of the public for public administration. *Administration and Society, 22*(4), 395–417.
Funtowicz, S. O., & Ravetz, J. R. (1993). Science for the post-normal age. *Futures, 25*(7), 739–755.
Geuijen, K. (2004). *De Asielcontroverse: Argumenteren over Mensenrechten En Nationale Belangen (The Asylum Controversy: Arguing about Human Rights and National Interests)*. Dutch University Press.
Geuijen, K., Moore, M., Cederquist, A., Ronning, R., & Van Twist, M. (2017). Creating public value in global wicked problems. *Public Management Review, 19*(5), 621–639.
Greely, H. (2020). Referenced in proceedings of National Academies of Sciences, Engineering, and Medicine. Solar geoengineering research governance: Proceedings of a workshop—in brief.
Gupta, A., Möller, I., Biermann, F., Jinnah, S., Kashwan, P., Mathur, V.,... & Nicholson, S. (2020). Anticipatory governance of solar geoengineering: conflicting visions of the future and their links to governance proposals. *Current Opinion in Environmental Sustainability, 45*, 10–19.
Jasanoff, S. (2020). National academies of sciences, engineering, and medicine. In *Solar Geoengineering Research Governance: Proceedings of a Workshop—in Brief*.
Kiser, L., & Ostrom, E. (1982). The three worlds of action: A metatheoretical synthesis of institutional approaches in strategies of political inquiry. *Strategies of Political Inquiry*, 179–222.
Lasswell, H. D. (1986). *Psychopathology and politics*. University of Chicago Press.
McGinnis, M. D. (2011). Updated guide to IAD and the language of the Ostrom workshop: A simplified overview of a complex framework for the analysis of institutions and their development. *Policy Studies Journal, 39*(1), 169–183.
Meynhardt, T. (2009). Public value inside: What is public value creation? *Intl Journal of Public Administration, 32*(3–4), 192–219.
Moore, M. H. (1997). *Creating public value: Strategic management in government*. Harvard University Press.
Oksanen, A. A. (2023). Dimming the midnight sun? Implications of the Sámi Council's intervention against the SCoPEx project. *Frontiers in Climate, 5*, Article 994193.
Ostrom, E. (2005). *Understanding institutional diversity*. Princeton University Press.
Ostrom, E. (2011). Background on the institutional analysis and development framework. *Policy Studies Journal, 39*(1), 7–27.
Ostrom, E. (2019). Institutional rational choice: An assessment of the institutional analysis and development framework. In *Theories of the policy process* (2nd ed., pp. 21–64). Routledge.
Ostrom, V., Tiebout, C. M., & Warren, R. (1961). The organization of government in metropolitan areas: A theoretical inquiry. *American Political Science Review, 55*(4), 831–842.

Paschke, M., Pfisterer, A., Backhaus, J., de Carteret, R., Damerius, L., Huang, Y., & Wallimann-Helmer, I. (2019). Applying collective inquiry. *Engaging in the science-policy dialogue* (Vol. 8, p. 10).

Pidgeon, N., Parkhill, K., Corner, A., & Vaughan, N. (2013). Deliberating stratospheric aerosols for climate geoengineering and the SPICE project. *Nature Climate Change, 3*(5), 451–457.

Rayner, S., Heyward, C., Kruger, T., Pidgeon, N., Redgwell, C., & Savulescu, J. (2013). The Oxford principles. *Climatic Change, 121*, 499–512.

Rosa, H. (2019). *Resonance: A sociology of our relationship to the world*. John Wiley & Sons.

Schmidt, V. A. (2008). Discursive institutionalism: The explanatory power of ideas and discourse. *Annual Review of Political Science, 11*, 303–326.

Schwartz, S. H. (2012). An overview of the Schwartz theory of basic values. *Online Readings in Psychology and Culture, 2*(1), 11.

Schwartz, S. H., Cieciuch, J., Vecchione, M., Davidov, E., Fischer, R., Beierlein, C., & Konty, M. (2012). Refining the theory of basic individual values. *Journal of Personality and Social Psychology, 103*(4), 663.

Whaley, L., & Weatherhead, E. K. (2014). An integrated approach to analyzing (adaptive) comanagement using the "politicized" IAD framework. *Ecology and Society, 19*(1).

Wylie, R. (Ed.). (2021). *Public value management institutional design and decision for the common good*. Rowman and Littlefield.

Public Engagement with Geoengineering: Emerging Technologies, Experimentation and the Climate Crisis

António Carvalho

Abstract This chapter delves into current discussions surrounding public engagement with geoengineering, examining the set of methodological and epistemological challenges posed by these emerging technologies to public participation with science and technology, as well as presenting some of the recurring trends stemming from these exercises. The chapter draws on empirical data following a research project on the Anthropocene in Portugal, which included six focus groups and a deliberative forum on carbon dioxide removal and solar radiation management with "situated" publics. The research design underlying these exercises was meant to "politicize" GE debates, emphasizing the performativity and ontological capacity of participatory devices. In the conclusion, the paper calls for the institutionalization of public engagement with GE; the development of a People's Charter for GE; the enactment of innovative forms of public engagement, arguing that public participation with GE should be taken seriously by decision-makers.

Keywords Public engagement · Participatory devices · Solar radiation management · Carbon dioxide removal · Focus groups · Climate crisis

1 Introduction

The aim of this chapter is to critically engage with the topic of public engagement with GE. Within the social sciences and environmental humanities, GE has stirred up much controversy. Swyngedouw and Ernston (2018) situate GE as an expression of the ecological modernization of capitalism, as it may represent the expansion of hegemonic socioeconomic and technological systems into new frontiers.

Szerszynski et al. (2013) argue that stratospheric aerosol injection (SAI), a specific form of GE, is problematic from a political perspective—drawing on Langdon Winner's reflections on the "politics of artifacts" (Winner, 1980) -namely the comparison between the politics of solar and nuclear energy—they argue that SAI is highly

A. Carvalho (✉)
Faculty of Economics and Centre for Social Studies, University of Coimbra, Coimbra, Portugal
e-mail: amcarvalho@ces.uc.pt

K.-U. Schrogl et al. (eds.), *The Politics of Geoengineering*, Springer Climate,
https://doi.org/10.1007/978-3-032-10095-5_12

undemocratic, representing a serious political danger. Others have argued that GE can be understood as an example of the many technofixes put forward to deal with the negative externalities of the Capitalocene, ignoring the socioeconomic roots of the current climate crisis (Altvater, 2016).

Although it has been suggested that social scientists are often more resistant to technofixes (such as geoengineering) than to top-down political solutions to tackle the climate crisis (such as the European Green Deal) (Heyward & Rayner, 2015), even its promoters recognize that GE raises a wide range of ethical and political issues (Crutzen, 2006), which for instance led to the development of the five Oxford GE principles.[1] GE research has also fueled conspiracy theories around "chemtrails" (Tingley & Wagner, 2017), which has inadvertently contributed to the public resistance toward the development of these technologies (Cairns, 2016). As we will see in this paper, due to its experimental character GE often problematizes the boundaries between present and future and fact and fiction, which presents additional problems to the task of identifying potential ethical, social and cultural implications of these technologies, as well as their health and environmental risks.

Instead of delving into the political, ethical and cultural implications of GE, this chapter examines some of the challenges it poses for public engagement and participation, considering the role of participatory devices for current democracies. After this introduction, the text delves into current scholarship on public participation and engagement with science and technology, including emerging technologies and GE, presenting some of the critical aspects. The third section delves into empirical data stemming from a series of focus groups on solar radiation management and carbon dioxide removal conducted at the University of Coimbra, Portugal, in December 2020, followed by a discussion which unpacks some of the tensions, contradictions and promises of public engagement with GE; the conclusion presents some of the limitations of the focus groups and lays out a series of recommendations regarding public engagement with GE.

2 Geoengineering and the Politics of Public Engagement with Science and Technology

Callon et al. argued that current democracies are characterized by a process of "double delegation", i.e., the delegation of political power to elected politicians and public officers and the delegation of decision-making in technical affairs to experts and scientists (Callon et al., 2011). In order to counter this scenario, they point toward a

[1] http://www.geoengineering.ox.ac.uk/www.geoengineering.ox.ac.uk/oxford-principles/principles/ The five principles are: Geoengineering to be regulated as a public good; Public participation in geoengineering decision-making; Disclosure of geoengineering research and open publication of results; Independent assessment of impacts; Governance before deployment.

wide range of "hybrid forums", participatory devices where citizens can have a say—and ideally influence—technological and scientific decision-making, often bringing together scientists and citizens.

The tension entwined with the apparent divide between "science" and "society" is also connected to a wider scholarly concern with citizen participation within western democracies, which is present at least since the 1960s (Arnstein, 1969). Regarding matters of science and technology, over the past 30 years a wide range of devices have been put forward to ideally overcome this divide, including focus groups, citizen forums, science shops, participatory assessment of technologies and even forms of citizen science (Carvalho & Nunes, 2018). Sociotechnical controversies such as nuclear energy, GMOs, biotechnologies and, more recently, nanotechnologies and GE, have highlighted the need to better engage citizens in the assessment—and even development—of technologies with potential social, ethical and even cultural impacts.

Within the European context, public engagement—and, recently, Responsible Research and Innovation—has played a central role in science and technology funding schemes, including Horizon2020 (the EU's research funding program between 2014 and 2020). This has highlighted the need to include the public in issues of science and technology, not only through science communication and education initiatives but also through what was coined "upstream engagement" (Carvalho & Nunes, 2018; Wilsdon & Willis, 2004), i.e., the engagement of "lay" publics while science and technology research was carried out, and not exclusively after it was conducted (downstream).

Nanotechnologies were considered a blueprint—and an opportunity—for the development of upstream forms of engagement (Rogers-Hayden et al., 2007). The research project DEEPEN—Deepening Ethical Engagement and Participation in Emerging Technologies (2006–2009), funded by the European Commission, carried out public engagement with nanotechnologies in Portugal and the United Kingdom, including focus groups and deliberative forums. This project integrated a methodological concern with the construction of "publics", relying on approaches linked to Paulo Freire's "Pedagogy of the Oppressed" and Augusto Boal's "Theater of the Oppressed" (Carvalho & Nunes, 2013, 2018).

DEEPEN's methodological concerns mirrored wider debates within science and technology studies with the "politics" of participatory devices. Some authors argued that public participation with science and technology was a way of legitimizing top-down decision-making in technical matters (Wynne, 2006), often reproducing the so-called "deficit model" of public understanding of science (Sturgis & Allum, 2004), which postulates that public resistance to specific innovations is due to the lack of "lay" knowledge on the topic. Consequently, participation—and especially "education"—would be a way of countering such resistance.

Others have recognized the role of participatory devices in the creation of publics (Felt & Fochler, 2010; Irwin, 2001), emphasizing how scholars—especially social scientists—contribute to the enactment of public perceptions of sociotechnical innovations and "matters of concern" (Latour, 2004). This was entwined with the "ontological turn" in science and technology studies (Pickering, 2017), which drew on notions such as performativity and ontological politics (Mol, 1999) to emphasize the productive dimension of participatory devices—instead of representing pre-existing "publics" and "concerns", they play an active role in the enactment of specific participatory realities, shedding light on the importance of methodologies, materialities, styles of moderation and institutional spaces of participation (Carvalho & Nunes, 2018).

Public engagement with emerging technologies such as nanotechnologies and synthetic biology was characterized by its experimental character, especially since these were still innovations in the making, with fluid boundaries between reality and (science) fiction (Schuijer et al., 2021). While, on the one hand, this may question the real-life impact of participatory exercises, on the other hand it paved the way for innovative forms of public engagement, often by resorting to performance, the arts and speculation, turning participatory devices into ontological events.

This brief contextualization was meant to present some of the historical and theoretical dynamics and tensions of public engagement with science and technology, which inevitably affects public engagement with GE. In 2009 the Royal Society published an impactful report, titled "Geoengineering the climate", which recognized the importance of public engagement with these techniques:

> The consequences of geoengineering—known and unknown, intended and unintended—would be felt by people and communities across the world. As with other emerging technologies, public participation in the development of research, governance and policy frameworks will be critical. (Royal Society, 2009: 42)

Similarly, the previously mentioned Oxford Geoengineering Principles also postulate the need for public participation as well as the disclosure of GE research, and in 2011 the IPCC conducted an Expert Meeting on Geoengineering in Lima, Peru, which included discussions on governance and social aspects of carbon dioxide removal and solar radiation management. Given the controversial character of these techniques, and a broader political context which emphasizes the need to engage citizens in the discussion of emerging technologies, over the past 10 years several initiatives of public engagement with GE have been carried out, including focus groups, qualitative workshops and surveys (Burns et al., 2016). Most of these initiatives have taken place in the Global North, especially in Europe and the US, where participatory mechanisms in science and technology are relatively well established.

Some of the recurring trends of public engagement with GE include, for instance, a concern with the "naturalness" of GE solutions; a preference for individual, social and political changes; the distinction between research and its large scale implementation; more public support for CDR than SRM; lack of knowledge about GE; and concerns about the democratic character of these applications (Burns et al., 2016; Corner et al., 2012; Merk et al., 2015). Some of these exercises include an experimentation with alternative forms of deliberation (Bellamy et al., 2017), namely art-based research and even attempts to explicitly politicize GE (Carvalho & Riquito, 2022), reinforcing the impact of broader theoretical and methodological discussions within science and technology studies on public engagement with these techniques.

Public engagement with GE is marked by some critical aspects. First, since these are applications still in the making, and the most spectacular versions of GE probably belong to the realm of fiction and imagination (such as solar mirrors), it is frequently problematic to determine what exactly participants are deliberating on, suggesting that GE is often too ambiguous to allow for a clear identification of its potential risks and impacts. Second, and although some exercises include scientists with knowledge about GE research, it is often the case that public engagement is led by social and human scientists with limited technical knowledge about these topics, which inevitably diminishes the potential impact of these exercises. Third, public engagement with GE is not "binding"—although some exercises may involve deliberative experiments potentially leading to written recommendations, these generally do not affect decision-making and technological innovation, which has fuelled previous critiques of the "uselessness" of public engagement with science and technology. Fourth, the experimental character of some of the participatory devices with GE may be an opportunity to depart from the traditional model of deliberation, based on the rational argument, which often reinforces epistemological and discursive asymmetries between participants; however, their "experimental" character may also be a recognition of the inability of public engagement to actually influence decision-making and technological innovation, consisting instead of empty academic fabulations and exercises; finally, most of these participatory initiatives take place in the Global North, ignoring that the Global South not only disproportionately pays the price of the climate crisis but also of the wide range of "technofixes" put in place to address it—including the extraction of rare earth minerals (Canelas & Carvalho, 2023). These critical points inevitably influenced the way in which the exercises on public engagement with GE were put in place in Portugal, as we will see in the following section.

3 Public Engagement with Geoengineering in Practice

This section presents a case study of public engagement with GE, namely six focus groups on solar radiation management and carbon dioxide removal conducted in Portugal in December 2020, the first participatory exercises on the topic in this southern European country. The focus groups were carried out as part of the research project TROPO—Ontologies of the Anthropocene in Portugal—social movements, public policies and emerging technologies, funded by the Portuguese Foundation of Science and Technology, which I coordinated between 2018 and 2022. The aim of TROPO was to assess a wide range of options to tackle the climate crisis, including the Portuguese Transition Network (linked to the Transition Town Totnes movement), focused on local low-carbon transitions; national strategies to achieve carbon neutrality, specifically the Portuguese Carbon Neutrality Roadmap 2050, aligned with the European Green Deal; national environmental controversies linked with the energy transition—such as lithium mining and the construction of large solar power stations in southern Portugal; public participation with GE, the focus of this chapter—this included six focus groups in 2020 and also an experimental deliberative forum in 2021.[2]

Within this project, public engagement with GE was part of a broader concern with the ontologies and politics of the climate crisis, therefore these exercises were informed by a concern with potential depoliticized technofixes, mirroring social science scholarship on the topic (Carvalho et al., 2021). Moreover, and taking into account scholarship on the "performativity" and politics of participatory devices, the aim was not merely to assess pre-existing stances on GE but to actively co-construct these ethicalities by selecting particular participants and group profiles, ideally with a stake on the climate crisis and its governance. The focus groups were, in that sense, an attempt to look at GE as encapsulating the "politics" of the climate crisis, i.e., the way in which these technologies can unfold multiple tensions, options and contradictions linked to hegemonic top-down and technological responses to climate change.

The focus groups included different "profiles" of participants. Group I was composed of environmentalists, belonging to different NGOs, environmental movements and grassroots organizations, and were expected to be particularly interested in the topic of climate change and GE. Group II was meant to emphasize the topic of intergenerational justice (which was brought to public attention particularly by Greta Thunberg and the School Strike for Climate) and included seven university students. Group III brought together five activists from a wide range of fields, such as feminism, LGBTQ+ rights, animal rights, trade unions and student movements, and the aim was to explore potential political and socioeconomic impacts of GE. Group IV included seven experts in the field of science communication, with backgrounds close to the topic of GE, such as volcanology, Geostatistics and Climate Sciences, and participants were expected to share a positive vision of technoscientific progress and to emphasize issues of transparency and education related to these emerging technologies. Group V included practitioners and advocates of alternative and holistic therapies and was expected to display a neo-luddite stance on science and technology, eventually rejecting technofixes such as GE, preferring instead more "holistic" solutions. Group IV was meant to function as a "control group", with participants not necessarily engaged with a particular stance/vision on science and technology and on the climate crisis, and included members from a heterogenous demography, both in terms of gender, age, location and professional status.

The aim of the focus groups was not to recruit a neutral/disengaged version of the wider "lay" public but to explicitly draw on the life experiences and situationality of participants to bring forward mediated perceptions of GE. This was informed by previous participatory exercises on nanotechnologies carried out at the Centre for Social Studies of the University of Coimbra within the project DEEPEN, where the research team drew explicitly on lessons stemming from the Pedagogy and Theater of the Oppressed, which emphasize the methodological capacity of individual and collective situatedness to learn, interpret and transform the world (Boal, 1979; Freire, 1970).

[2] For the sake of this chapter the empirical data is centered exclusively on the focus groups.

The focus groups started off with a presentation on the climate crisis, followed by information on GE, specifically on the particularities of solar radiation management and carbon dioxide removal, including ongoing research and potential future applications (namely some spectacular ones such as space mirrors). The research team had a script with a series of topics to guide the discussion, including questions related to GE and the climate crisis; ethical issues; political aspects; daily impact of GE; specific aspects about Portugal; finally, the focus groups also included questions tailored for each specific group, such as for instance its impact on environmentalism, social movements and struggles, and how it affected science communication. In practice, the focus groups were conducted informally, without rigidly resorting to a script, to better allow topics and discussion to arise naturally. The focus groups were fully recorded and transcribed, and all transcripts were anonymized. All participants signed an informed consent form. The focus groups were conducted via Zoom due to the COVID-19 pandemic.

The focus group discussions have been extensively presented and analyzed in a previous publication (see Carvalho & Riquito, 2022). In this chapter, I present a summary of these findings, focusing on seven main topics. The first topic has to do with unexpected risks: most participants highlighted the level of uncertainty surrounding GE, namely negative environmental impacts, but also the risk of a "moral hazard" (Corner & Pidgeon, 2014), i.e., the possibility that GE would decrease the need for substantial changes in socioeconomic systems necessary to tackle the "roots" of the climate crisis, including behavioral and systemic changes. The second topic concerned uncertainty and unfamiliarity—in fact, it was the first time that most participants were hearing about GE (with the exception of science communicators and some environmental activists), which limited their ability to formulate potential ethical and social impacts of these technologies and even added a certain degree of ambiguity to GE itself, with some participants unable to understand what was really at stake (namely, if what was being debated were "technological" and "artificial" versions of GE, such as stratospheric aerosol injection, or natural approaches such as afforestation). The third topic concerned the idea of GE as a Band-aid—participants argued that GE was a technofix unable to solve the deep-rooted dimensions of the climate crisis, which included aspects such as: capitalism (as put by some activists and environmentalists); modern/dualist and masculinist engagements with the environment (a stance espoused by advocates of holistic/alternative therapies); and consumerism (some environmentalists advocated profound behavioral changes).

Participants argued that GE was not a "real solution" and therefore advocated for wider, systemic changes, often mentioning the risk that these emerging technologies could constitute a moral hazard preventing these necessary changes to take place.

The fourth topic had to do with the idea that GE perpetuated global inequalities, namely economic asymmetries between the North and the Global South, reminding us of current debates around climate coloniality (Sultana, 2022)—according to participants, not only could GE be appropriated by capitalist systems and turned into a "profit" to benefit large corporations but countries in the Global South—already the most affected by climate change—would also be disproportionately affected by the negative impacts of GE. Similarly, another relevant topic concerned the fact that GE was considered by some as an expansion of the capitalist and extractivist paradigm, reproducing the same economic system and ontological stance at the roots of the climate crisis. The sixth topic concerned GE as a climate justice matter, which raised issues dealing with colonialism, the role of local and indigenous communities and the participation of the wider public in climate policies and future technologies. Finally, the seventh trend concerned the relationship between GE and the Portuguese context, with participants relying on several environmental and sociotechnical controversies at the national scale to "build" their imaginaries of carbon dioxide removal and solar radiation management, arguing that the governance and unfolding of these controversies—such as lithium mining or Spanish nuclear stations close to Portuguese borders—and Portugal's semi-peripheral situation in Europe were relevant trends that would decisively shape the future impacts of GE in their country.

As we have seen, the focus groups were an explicit attempt to "politicize" GE debates, and it could be argued that the focus groups operated as a self-fulfilling prophecy—by recruiting "situated" participants, with a potential stake in the topic, the discussions were inevitably marked by a series of sociopolitical, cultural and economic tensions underpinning the climate crisis as well as by a wide range of technologies put forward to address it. In that sense, public engagement with GE was a way of shedding light on the politics of the climate crisis itself. It could be argued that the social and ethical concerns emerged in a performative way, in the sense that they were mediated by individual identities, group composition, and previous stances on other technologies and sociotechnical and environmental issues. GE becomes a "matter of concern" through the participatory device itself—in this case the focus groups—allowing citizens to discuss and deliberate on technologies in the making, ambiguous entities that, although characterized by uncertainty, become mirrors of wider environmental and sociotechnical phenomena.

Social scientists become mediators in the enactment of GE ethics, mobilizing material and discursive devices which perform matters of concern. Not only are matters of concern performatively enacted but GE itself is also "constructed" and presented in certain ways, which highlights the epistemological capacity of participatory devices, as they play a role in the creation of public perceptions of these technologies, inevitably shaping the way in which civil society perceives these innovations.

The performativity of the focus groups and of public engagement in general is also an invitation to question what is the "public" of public engagement (Bellamy & Lezaun, 2017). As we have seen in this section, the public was selectively crafted in order to trigger a discussion on the "politics" of GE. This suggests that particular situationalities can be ways of unpacking particular matters of concern, which requires social scientists—and those in charge of organizing public engagement exercises with GE—to take into account distinct epistemological, cultural and ethnic backgrounds while debating emerging technologies, thus avoiding simplistic generalizations of a pre-situated and neutral "lay public".

4 Conclusion and Recommendations

The focus groups were an interesting case study to unpack the dynamics of public engagement with GE in practice, articulating some of the concepts and tensions explored in section II. However, they were marked by a number of limitations: first, the focus groups were not binding, and although some participants formulated recommendations on GE governance, the main concern was to debate ethical and sociopolitical implications of solar radiation management and carbon dioxide removal and not necessarily to influence decision-making, research and implementation; second, scientists (especially those involved in GE research) were not directly involved in the preparation and conduction of the focus groups, which may have contributed to a caricatural and biased view of these technologies, as they were "translated" by social scientists not formally trained in research fields relevant to their development; third, this was a relatively small sample of "engaged" and "situated" participants, which may not represent the Portuguese population as a whole—consequently, the "politicized" version of GE that emerged during the discussions should be taken with a grain of salt, as other situated populations—in distinct national, cultural and geographical contexts—may enact rather different "matters of concern".

Due to the nature of this book, some tentative recommendations concerning public engagement with geoengineering are discussed in the following paragraphs.

Public engagement with these technologies should ideally be institutionalized—currently, these exercises are often conducted within the scope of research projects in the field of social and human sciences, often in European and North American countries. This often fails to attain a global impact and rarely reaches those directly involved in decision-making and GE research. Since the climate crisis is a global phenomenon, and some GE applications may be hard to circumscribe to specific national boundaries (such as stratospheric aerosol injection), public engagement should also be conducted at the global scale, and should be institutionalized by bodies such as the United Nations, to ensure that citizens can take part in forms of upstream engagement with these techniques, ideally in a binding—and not merely ritualistic or tokenizing—way. Other national and transnational institutions—such as the European Commission—should also institutionalize these events, ensuring that GE research and implementation should not happen without citizen deliberation and that these processes are transparent, binding and inclusive.

Second, and although the Oxford principles are a good start when it comes to ethical recommendations regarding GE, it would be important to develop a People's Charter for Geoengineering which would outline in detail aspects related to research, governance and implementation of the most relevant techniques. During COP27, in Sharm El Sheikh, Egypt, there was an online Panel Discussion on the Ethics of Climate Engineering. In the near future there could be a deliberative forum GE organized by COP with the aim of producing binding recommendations on these technologies. This deliberative forum, with the aim of producing a People's Charter, would include citizens from all the five continents, and would entail presentations by leading GE researchers, to ensure that deliberation would be aligned with the state of the art. The charter would then be made public and could be included in a potential UN treaty on GE, to ideally regulate its present and future research and implementation.

Third, GE can be understood as an opportunity to develop innovative forms of public engagement. Online deliberation may allow those in the Global South to better inform decision-making processes and research in the Global North that may impact them (taking into account current debates around climate coloniality), as well as other subaltern populations. Currently, there are also concerns with more-than-human participation, driven in part by Bruno's Latour work on the Parliament of Things (Latour, 2004, 2017)—as GE will have an impact not only on human populations but also on Earth Systems, it will be important to develop mechanisms to assemble a wider community in processes of deliberation and political engagement, which requires innovative and experimental forms of engagement drawing on the arts, the senses and speculative research (Wilkie et al., 2017).

Fourth, as the climate crisis accelerates, solutions such as GE will probably become part of mainstream discussions. This will eventually trigger public mistrust due to the complex, futuristic and somehow dystopian characteristics of these techniques, making science communication and public engagement urgent. This means that not only climate change communication but also public engagement with geoengineering should be taken seriously by governing bodies—if citizens are ignored, or if participation is merely at the degree of tokenism (Arnstein, 1969) this will eventually generate public backlash and confrontation, inevitably undermining GE as a legitimate solution to address the climate crisis.

Acknowledgements I would like to thank Mariana Riquito, the research assistant of TROPO, for her support in the organization of the focus groups and the deliberative forum. This work is funded by ERDF—European Regional Development Fund through the COMPETE 2020—Operational Program for Competitiveness and Internationalization (POCI), and by Portuguese funds through FCT—Fundação para a Ciência e a Tecnologia, project TROPO, no. 028362 (POCI-01-0145-FEDER-028362).

References

Altvater, E. (2016). The capitalocene, or, geoengineering against capitalism's planetary boundaries. In J. Moore (Ed.), *Anthropocene or capitalocene* (pp. 138–152). PM Press.

Arnstein, S. R. (1969). A ladder of citizen participation. *Journal of the American Institute of Planners, 35*(4), 216–224.

Bellamy, R., & Lezaun, J. (2017). Crafting a public for geoengineering. *Public Understanding of Science, 26*(4), 402–417.

Bellamy, R., Lezaun, J., & Palmer, J. (2017). Public perceptions of geoengineering research governance: An experimental deliberative approach. *Global Environmental Change, 45*, 194–202.

Boal, A. (1979). *Theatre of the oppressed*. Pluto Press.

Burns, E. T., Flegal, J. A., Keith, D. W., Mahajan, A., Tingley, D., & Wagner, G. (2016). What do people think when they think about solar geoengineering? A review of empirical social science literature, and prospects for future research. *Earth's Future, 4*(11), 536–542.
Cairns, R. (2016). Climates of suspicion: 'chemtrail' conspiracy narratives and the international politics of geoengineering. *The Geographical Journal, 182*(1), 70–84.
Callon, M., Lascoumes, P., & Barthe, Y. (2011). *Acting in an uncertain world: An essay on technical democracy*. MIT Press.
Canelas, J., & Carvalho, A. (2023). The dark side of the energy transition: Extractivist violence, energy (In)justice and lithium mining in Portugal. *Energy Research and Social Science, 100*, 103096.
Carvalho, A., Ferreira, V., & Matos, A. R. (2021). Ontologias do Antropoceno: Crise climática, respostas sociopolíticas e tecnologias emergentes. *Forum Sociológico Série II*, 5–13.
Carvalho, A., & Nunes, J. A. (2013). Technology, methodology and intervention: Performing nanoethics in Portugal. *NanoEthics, 7*, 149–160.
Carvalho, A., & Nunes, J. A. (2018). Assembling upstream engagement: The case of the Portuguese deliberative forum on nanotechnologies. *NanoEthics, 12*, 99–113.
Carvalho, A., & Riquito, M. (2022). It's just a Band-Aid!': Public engagement with geoengineering and the politics of the climate crisis. *Public Understanding of Science, 31*(7), 903–920.
Corner, A., & Pidgeon, N. (2014). Geoengineering, climate change scepticism and the 'moral hazard' argument: An experimental study of UK public perceptions. *Philosophical Transactions of the Royal Society a: Mathematical, Physical and Engineering Sciences, 372*(2031), 20140063.
Corner, A., Pidgeon, N., & Parkhill, K. (2012). Perceptions of geoengineering: Public attitudes, stakeholder perspectives & the challenge of 'upstream' engagement. *Wiley Interdisciplinary Review of Climate Change, 3*, 451–466.
Crutzen, P. J. (2006). Albedo enhancement by stratospheric sulfur injections: A contribution to resolve a policy dilemma? *Climatic Change, 77*(3–4), 211.
Felt, U., & Fochler, M. (2010). Machineries for making publics: Inscribing and de-scribing publics in public engagement. *Minerva, 48*(3), 219–238.
Freire, P. (1970). *Pedagogy of the oppressed*. Continuum.
Heyward, C., & Rayner, S. (2015). Uneasy expertise: Geoengineering, social science, and democracy in the Anthropocene. In M. Heazle & J. Kane (Eds.), *Policy legitimacy, science and political authority* (pp. 101–121). Routledge.
Irwin, A. (2001). Constructing the scientific citizen: Science and democracy in the biosciences. *Public Understanding of Science, 10*(1), 1–18.
Latour, B. (2004). *Politics of nature*. Harvard University Press.
Latour, B. (2017). *Facing Gaia: Eight lectures on the new climatic regime*. John Wiley & Sons.
Merk, C., Pönitzsch, G., Kniebes, C., Rehdanz, K., & Schmidt, U. (2015). Exploring public perceptions of stratospheric sulfate injection. *Climate Change, 130*(2), 299–312.
Mol, A. (1999). Ontological politics. A word and some questions. *The Sociological Review, 47*(1), 74–89. https://doi.org/10.1111/j.1467-954X.1999.tb03483.x
Pickering, A. (2017). The ontological turn: Taking different worlds seriously. *Social Analysis, 61*(2), 134–150. https://doi.org/10.3167/sa.2017.610209
Rogers-Hayden, T., Mohr, A., & Pidgeon, N. (2007). Introduction: Engaging with nanotechnologies–engaging differently? *NanoEthics, 1*, 123–130.
Royal Society. (2009). *Geoengineering the climate: Science, governance and uncertainty*. Royal Society Policy Document.
Schuijer, J. W., Broerse, J. E., & Kupper, F. (2021). Citizen science fiction: The potential of situated speculative prototyping for public engagement on emerging technologies. *NanoEthics, 15*, 1–18.
Sturgis, P., & Allum, N. (2004). Science in society: Re-evaluating the deficit model of public attitudes. *Public Understanding of Science, 13*(1), 55–74. https://doi.org/10.1177/0963662504044432690
Sultana, F. (2022). The unbearable heaviness of climate coloniality. *Political Geography, 99*, Article 102638. https://doi.org/10.1016/j.polgeo.2022.102638

Swyngedouw, E., & Ernstson, H. (2018). Interrupting the Anthropo-obScene: Immuno-biopolitics and depoliticizing ontologies in the Anthropocene. *Theory, Culture & Society, 35*(6), 3–30.

Szerszynski, B., Kearnes, M., Macnaghten, P., Owen, R., & Stilgoe, J. (2013). Why solar radiation management geoengineering and democracy won't mix. *Environment and Planning A, 45*(12), 2809–2816.

Tingley, D., & Wagner, G. (2017). Solar geoengineering and the chemtrails conspiracy on social media. *Palgrave Communications, 3*(1), 1–7.

Wilkie, A., Savransky, M., & Rosengarten, M. (Eds.). (2017). *Speculative research: The lure of possible futures* (1st ed.). Routledge. https://doi.org/10.4324/9781315541860

Wilsdon, J., & Willis, R. (2004). *See-through science: Why public engagement needs to move upstream.* Demos.

Winner, L. (1980). Do artifacts have politics? *Daedalus, 109*(1), 121–136. https://www.jstor.org/stable/20024652

Wynne, B. (2006). Public engagement as a means of restoring public trust in science: Hitting the notes, but missing the music? *Community Genetics, 9*(3), 211–220.

GE Extended: Terraforming on Celestial Bodies

Chris Pak

Abstract Science fiction (sf) analyses and critiques terraforming by situating theoretical and speculative ideas within wider social, historical and conceptual contexts and by narrating the consequences of specific decisions or frameworks for adapting and colonising celestial bodies. Sf can be used as an experimental forum for developing and scrutinising strategies for projects of planetary dimension. Terraforming sf engages debate about deliberation and the possibility for developing international frameworks for decision making and co-ordination under crisis conditions. This chapter considers Kim Stanley Robinson's Mars trilogy, Jane Killick's In the Shadow of Deimos: A Terraforming Mars Novel, Mary Robinette Kowal's Lady Astronaut trilogy, Andy Weir's Artemis and Project Hail Mary and Ian McDonald's Luna trilogy to consider the implications of catastrophe on decision making, the emergence of new institutions that make terraforming possible, the corporate exploitation of space and the importance of the formation of new identities for the persistence of interplanetary colonies.

Keywords Geoengineering · Terraforming · Science fiction · Governance · Identity · Space

1 Introduction

Science fiction (sf) can be used as an archive of scenarios for thinking through the socio-technical, cultural, political and material transformations that developments in science and technology could create (Milner & Burgmann, 2020, p. 27; Boyer & Szeman, 2014; Butler 2014; Milner, 2012, pp. 180–181; Steinmüller, 2003). This chapter reads sf terraforming stories to think through issues of policy and governance related to the scientific, social and cultural value of the solar system, as well as its commercial exploitation. Sf as a body of fiction has been characterised as an sf megatext, within which individual works not only rely on predecessors that have

C. Pak (✉)
Swansea University, Wales, UK
e-mail: c.a.pak@swansea.ac.uk

K.-U. Schrogl et al. (eds.), *The Politics of Geoengineering*, Springer Climate,
https://doi.org/10.1007/978-3-032-10095-5_13

helped to construct popular ideas about speculative technologies and phenomena, but also against which they respond to the scenarios portrayed in earlier fictions, sometimes reaffirming and at others revising, updating and challenging the visions of the future that they portray (Attebery & Hollinger, 2013; Morrissey, 2000; Seed, 2000; Broderick, 1995).

Sf is particularly suited to exploring issues related to terraforming: sf writer Jack Williamson (2004) coined 'terraforming' in his 1942 short story 'Collision Orbit' and throughout the twentieth and twenty-first centuries other writers have continued to portray a range of approaches to the terraforming of celestial bodies that are expressive of different orientations to governance, policy and identity. The value of such fictions for providing speculative analyses and critiques of approaches to terraforming lie in their capacity to situate theoretical and speculative ideas within wider social, historical and conceptual contexts and their ability to narrate the consequences of specific decisions or frameworks relating to the adaptation and colonisation of celestial bodies.

Sf connects the terraforming of celestial bodies to geoengineering on Earth in multiple ways. Sf positions Earth and its interplanetary colonies as mirrors of one another to explore orientations and stances to the management of environments, to offer fictional experiments in technical and social reconfiguration and to examine actors' escalating abilities to shape space. Sf can offer insights into the public imaginary about how the future of terraforming has been conceived, and it can thus help us to understand public orientations to terraforming and the uses of space.

The timescale imagined for terraforming positions this activity in the far future. Terraforming will require significant resources, modes of social, economic and political organisation, and technical capacities that are beyond current capabilities. This timescale means that engagement with terraforming can function as an experimental forum for developing strategies for projects of planetary dimension. As the repercussions of terraforming are displaced in a future that is all but inaccessible reflection on terraforming can function as a substitute for reflection about strategies for addressing climate change on Earth and for conceptions of a green transition, which could offer opportunities for public engagement that circumvents eco-anxiety or which speaks directly to climate change.

Sf's key contribution to debates about terraforming centre on its ability to narrate the contexts within which national and international approaches to framing terraforming and the uses of space are conceived. The implications of metaphors such as the conquest of space, space as a frontier and Earth as humankind's cradle imply a series of relationships to space and terraforming that need to be addressed. Indeed, Carl Iain Cater (2014, p. 839) argues that 'contemporary developments in extraterrestrial tourism are actually just an extension of the "frontier culture."' The large-scale planetary endeavour that terraforming represents will require mechanisms for public deliberation and international consensus about the purposes and modes of adaptation that are imagined as well as frameworks for the private development and exploitation of potential resources. The cultural value of space and the potentially traumatic transformations to celestial bodies that have significant and longstanding value for different communities and cultures on Earth will also condition how terraforming will be conducted.

Furthermore, appropriate modes of engagement with nature and the possibility of encountering life will be key aspects of terraforming policy, but so will an appropriate environmental ethics for seeding life on other planets and adapting existing planetary environments. A wider infrastructure will also be required, and thus this infrastructure needs to be assessed for how well it affords equitable access to space and for its alignment with an environmental-ethical policy. International frameworks for space tourism, the management of space junk, conflict, commercialisation and an energy policy are also key dimensions to speculation about terraforming. To address these questions this chapter will consider works of sf such as Kim Stanley Robinson's (1996a, 1996b, 1996c) landmark Mars trilogy (comprising Red Mars, Green Mars and Blue Mars), Jane Killick's (2021) In the Shadow of Deimos: A Terraforming Mars Novel, Mary Robinette Kowal's (2013, 2019a, 2019b, 2020) Lady Astronaut trilogy (comprising the novelette 'The Lady Astronaut of Mars' and the trilogy proper, The Calculating Stars, The Fated Sky and The Relentless Moon), Andy Weir's (2017, 2021) Artemis and Project Hail Mary and Ian McDonald's (2015, 2017, 2019) Luna trilogy (comprising Luna: New Moon, Luna: Wolf Moon and Luna: Moon Rising).

2 Deliberation and International Frameworks Under Crisis Conditions

In Weir's (2021) Project Hail Mary, a microscopic alien lifeform dubbed 'astrophage' is discovered, which feeds on the sun's light and thus intercepts it before it reaches Earth. While Project Hail Mary does not appear to be about terraforming its depiction of alien life that is able to shield the Earth from the sun's radiation positions it as a form of terraforming. Extrapolation of the effects of its siphoning of the sun's energy leads to the discovery that should this reduction continue, the Earth will undergo a disastrous global cooling. Astrophage's novelty and the urgency of the situation compel Earth's nations to establish a state of exception and to collaborate to develop and implement solutions to this problem.

International efforts to establish global research networks and to provide other scientific institutions and researchers with access to the astrophage samples serve as an analogy that outlines the potential scale of terraforming projects and establishes one scenario for thinking about the complexities of the endeavour. Project Hail Mary provides insight into how normative frameworks for international co-operation may be superseded under emergency conditions and can be used to reflect on some of the challenges relating to international co-ordination and sovereignty that terraforming introduces, should such measures be framed as a response to climate change, over-population and ecological and societal collapse on Earth. Catastrophe serves as a justification for the suspension of international processes for decision making. Eva Stratt, in perhaps one of the most incredible of science-fictional premises evidenced in the sf terraforming and geoengineering literature, has by national consensus been given supreme authority over the Earth's nations to facilitate a response to the disaster. She explains that 'There is no time for slow, international committees' (Weir, 2021, 39). The speed with which Stratt assembles her task force and the overarching powers she is granted to compel private individuals and national actors to develop solutions to the catastrophe creates conflict, for which Stratt leverages the threat of force provided by the US army. The solutions deemed viable within the timeframe for meaningful response results in exclusions from protection for anything other than humans. 'I don't care about morality,' Stratt says in response to one solution to delay a rapid cooling by bombing the Antarctic to emit methane into Earth's atmosphere (Weir, 2021, 235). Another solution, to convert the Sahara into a factory for generating energy using the astrophage itself, 'would destroy the ecology of Africa and probably Europe' (Weir, 2021, 220).

The problems associated with international frameworks that are too slow to respond to global catastrophe is a key issue for terraforming, especially when one rationale for terraforming, raised in Robinson's (1996a, 1996b, 1996c) Mars and Kowal's (2013, 2019a, 2019b, 2020) Lady Astronaut trilogies, is that it offers a safety valve for populations facing the repercussions of Earth's changing climate. As Weir's (2021, 391) protagonist explains, 'Right now we're soft. You, me, the whole Western world. We're the result of growing up in unprecedented comfort and stability.' Deliberation is seen in this story as a symptom of this softness, thus framing stability as constitutive of the difficulties for international co-operation on global and interplanetary scale projects. Project Hail Mary identifies moral considerations for non-humans, increased wellbeing and the observance of precautionary principles as key problems for emergency response.

This view of deliberation in the context of catastrophe has a long history in sf stories of terraforming. Robinson's (1996a, 1996b, 1996c) Mars trilogy includes scenes in which deliberation is circumvented, as when the chief terraformer Sax Russell initiates several projects to terraform Mars in Red Mars without seeking a broader consensus among the other scientists of the community. This failure to seek broad-based consensus is indicative of initial approaches to decision making among the burgeoning Martian community and speaks to the growing conflict and eventual schism between those who support terraforming by any means necessary and those who desire a slower approach. Russell's initial disdain for deliberation threatens to reinforce pre-existing differences between supporters and detractors of the project if his interventions were to be made widely known. Deliberation is a key problem and site of conflict throughout the Mars trilogy and differences between how Mars should be terraformed—differences held by those on Earth as much as between those on Mars—drive much of the story, particularly in the final instalment of the trilogy, Blue Mars. In this novel, a newly independent Mars seeks to develop through sustained public discourse a Martian constitution that would codify Martian independence and establish an identity capable of embracing all of the Martian communities.

If Weir's (2021) Project Hail Mary identifies deliberation as a key problem and site of conflict, then Robinson's (1996a, 1996b, 1996c) Mars identifies its lack as a danger attending terraforming. In the Mars trilogy, climate change on Earth and dwindling natural resources establish a sense of urgency that motivates Mars colonisation, though the stakes involved are not presented as starkly as in Weir's (2021) novel. Kowal's (2013, 2019a, 2019b, 2020) Lady Astronaut trilogy, like Weir's (2021) Project Hail Mary, is framed against the background of a runaway climate catastrophe that resembles in scope that which Weir presents. In this alternate 1950s US, an asteroid collides with the Eastern US states, killing the president and much of the highest-ranking political class. Missions to the Moon and Mars to establish colonies in the event that Earth needs to be evacuated are put in place. Much of the trilogy explores the prejudices that women, people of colour and broader swathes of the US population who are excluded from or involved in the colonising project face, despite the sorely needed technical expertise and knowledge that they offer to the project. However, broader dissent emerges among those who view the space missions as exclusionary, who understand that the moon and Mars cannot offer sanctuary for the whole of the Earth's population and those who wish that the resources spent on the project would be used to tackle climate change on Earth. Thus, issues of appropriate modes of deliberation among a wider populace form a key aspect of this trilogy's engagement with politics, while the historical exclusion of non-white, non-male stakeholders are shown to be key issues affecting terraforming projects.

3 Institutions Involved in Terraforming

In the context of catastrophe and in acknowledgement of the multiple interests that motivate terraforming, sf identifies new institutions or modifications to existing institutions as central to the colonisation of other worlds. In Killick's (2021) In the Shadow of Deimos, a world government on Earth has established Martian colonies to terraform the planet, which points to the forms of international co-ordination that terraforming requires. A United Nations Mars Initiative oversees terraforming but is undermined by the world government's invitation to private corporations to invest in and develop Mars in 2315. The terraforming committee is portrayed as uncritically welcoming of corporate development despite the controversial methods adopted by some of the most high-profile of chief executives and, indeed, much of the novel's plot centres on a corporate conspiracy to gain supremacy over key portions of the Martian land. Deimos focusses on the abuses that corporate actors, situated on Mars and thus subject to lax oversight, could engage when private investment drives terraforming. A key concern in this novel is how power fails to be distributed but rather is retained by the private corporations operating on Mars, who control all aspects of their worker-colonists' lives. Corporate oversight is consequently curtailed and avenues for employees to appeal to external bodies and to engage in whistle-blowing are unavailable in this system, leading one protagonist to circumvent legal frameworks to seek appropriate modes of justice.

In Weir's (2017, p. 37) Artemis, private investment is essential for establishing a lunar colony. In the novel's past former Kenyan minister of finance Fedelis Ngugi single-handedly created a national space industry by offering equatorial access for space launches and tax breaks to fifty corporations from thirty-four countries. Ngugi knowingly courts investment from criminal organisations who use their Lunar corporation to launder money from Earth through the unregulated slugs which serve as currency on Artemis. As Ngugi explains, "'Artemis wouldn't exist without syndicate money. Idealism doesn't put Gunk on people's plates'" (Weir, 2017, p. 178). The lack of appropriate legal frameworks and the cultivation of a criminal underbelly enables actors to operate with little oversight, thus making it attractive for investment. Following the model of the American frontier town, Ngugi explains that:

> It's all part of the life-cycle of an economy. First it's lawless capitalism until that starts to impede growth. Next comes regulation, law enforcement, and taxes. After that: public benefits and entitlements. Then, finally, over-expenditure and collapse. (Weir, 2017, p. 299)

While Artemis' primary economic bases are tourism and, more importantly, immigration, it is enabled by initial corporate investment that can only be attracted through tax cuts and, unofficially, lax oversight. Crucially, however, this largely unregulated economy must give way to new industries and regulatory frameworks if it is to persist. This model for the economic development of interplanetary colonies is based on an analogy between the settlement of the North American West and interplanetary colonies but is by no means the only model for colonisation that is available historically nor within the sf literature. Nevertheless, Artemis offers an environment for producing rare materials and technologies that would be impossible to create on Earth, and the novel turns on the development of one such technology that promises to reshape the telecommunications industry.

One risk associated with the uneven relationships between governmental bodies and private corporations is how private individuals and employees become victims to machinations for greater power on Mars. Another risk, however, is that corporate actors may engage in terraforming approaches that conflict with other strategies and which place populations at risk. This is the case in Killick's (2021) Deimos, in which novel a corporate plot involving the sabotage of a meteor to destroy a research facility leads to the unplanned death of a technician. As Deimos suggests, when the desires of private actors go unregulated, and when investment is allowed to drive terraforming agendas, abuses of power, corporate malfeasance and corruption are invited. Weir's (2017) Artemis, however, presents governmental bodies and private actors as collaborating to develop Lunar colonies by making use of the possibilities afforded by a lack of regulation and oversight.

In McDonald's (2015; 2017; 2019) Luna trilogy private corporations, based on family structures, exercise complete control over extraction and the development of the Moon. Contracts regulate how the ruling oligarchies conduct relations with one another and their employees. Indeed, so long as appropriate contracts are established then any form of conflict between Houses is sanctioned, including war, assassination and espionage. As with many stories of terraforming, from Robinson's (1996a, 1996b, 1996c) Mars trilogy and Weir's (2017) Artemis, back to some of the earliest terraforming stories such as Robert A. Heinlein's (2001) 1966 The Moon is a Harsh Mistress and Frederik Pohl and C.M. Kornbluth's (1974) 1953 The Space Merchants, governmental and corporate actors position terraformed planets as sites for the extraction of resources for Earth. The colonised planet's status as an annexe for the supply of materials to Earth introduces another actor, that of the often newly independent colony, which seeks to establish self-determination and economic and political independence from Earth.

In Deimos the terraformed planet is firmly positioned as subject to governance from Earth, with a terraforming council managing interactions between corporate actors and between the world government and Mars, and thus independence does not factor into the narrative. In Robinson's (1996a, 1996b, 1996c) Mars trilogy, however, rejection of the status of Mars as a locale for extraction leads Mars to establish, through a failed civil war, years of resistance and a global-scale war for independence, a new national identity embracing the whole of the planet. The nations on Earth, corporate actors on Mars and the Martian communities, represented by a government of its own in Blue Mars, are key actors in this vision of terraforming.

Stories of Martian independence abound and suggest that colonies cannot be governed across the extremes of space from Earth. These stories are informed by histories of the colonisation of the North American West and thus raise analogies between American history and the relationships that may coalesce on other planets. While stories such as Kowal's (2013, 2019a, 2019b, 2020) Lady Astronaut trilogy firmly position terraforming as led by American interests, others such as McDonald's (2015; 2017; 2019) Luna trilogy show how private actors working beyond national frameworks will drive terraforming for their own interests. Still others, such as Robinson's (2018) Red Moon and Weir's (2017) Artemis, suggest that multiple national actors will collaborate or work independently—and sometimes at cross-purposes—to develop colonies or multiple enclaves on other planets. Interaction and co-ordination between enclaves will thus be of central importance, and many private actors are positioned as key to developing infrastructures on other planets. In Robinson's (1996b) Green Mars, for instance, corporations such as the fictional Praxis develop diplomatic relations with the independent Mars communities prior to the establishment of full economic and political independence from Earth. Praxis seeks to work with the colonists to establish new economic models for private investment in Mars and for realising the benefits of terraforming research and management on Earth. This alternate corporate strategy depends on agents who exercise a diplomatic function. Art Randolph is tasked on behalf of Praxis to liaise with the various Martian communities to establish relations with Praxis. As Randolph lives and works on Mars, meeting a variety of communities, corporate representatives and de facto leaders, he commits to Martian independence. Rather than viewing this commitment as an abdication of his responsibilities to Praxis, CEO William Fort sees Randolph's investment as a natural consequence of his work and an extension of his diplomatic mission to Mars. Commitment to the Martian communities is thus central to this view of the relationship between private actors, the Mars communities and governmental actors on Earth and Mars.

Other institutions include a global court in Blue Mars to regulate terraforming and to ensure that any project conforms to the environmental constraints established in the Martian constitution. Such systems echo the world government on Earth in Deimos and the courts of McDonald's (2015; 2017; 2019) Luna trilogy, and raises questions about co-ordination, management and control. Global perspectives are often viewed as essential for terraforming as such endeavours necessarily impact on a planet's global parameters, thus highlighting similar issues and potential conflicts related to geoengineering on Earth. In Mars, these methods of co-ordination and legal frameworks for environmental-ethical observation are essential to the new identities that coalesce on Mars.

4 Insights from Sf About Terraforming Governance and Policy

The sf literature on terraforming identifies the status of the terraformed body and its relationship to actors on Earth as critical to broad-based acceptance of terraforming and its viability. Terraformed planets offer the potential for applied research and development and foundational scientific study, as Ann Clayborne argues in Robinson's (1996c) Red Mars. However, the status of celestial bodies as sites for extraction can conflict with uses of terraformed planets for scientific research, as Brian Aldiss and Roger Penrose's (1999) White Mars suggests, in which novel the Martian colonists build a particle accelerator to conduct high-level research into particle physics. Unlike the modes of extraction depicted in Weir's (2017) Artemis, Killick's (2021) Deimos and McDonald's (2015; 2017; 2019) Luna trilogy, Robinson's (1996a, 1996b, 1996c) Mars trilogy suggests that policies regulating methods of extraction and its scope are fundamental to ensuring that other uses of the terraformed planet are satisfied. These policies seek to address potential forms of corporate exploitation of labour and land, which will have repercussions for the nature of the communities that develop on Mars. Indeed, as Robinson's (1996a, 1996b, 1996c) Mars trilogy suggests, strict economic specialisation on terraformed bodies risks narrowing the expertise, forms of cultural development and the value of Mars, which in turn will have corresponding implications for the longevity of terraforming projects.

Forms of global regulation are also required to co-ordinate terraforming approaches and to ameliorate the risk of strategies negating the effects of others. Thus, the forms of corporate terraforming depicted in Deimos introduce excessive risk of harm to people and the planet. However, as Robinson's (1996a, 1996b, 1996c) Mars trilogy shows, such global co-ordination cannot be regulated from Earth but should be managed by the various communities on celestial bodies undergoing terraforming. Regulatory bodies and legal processes must be situated on Mars as the distance between Earth and other celestial bodies will hamper efforts to monitor and adjudicate potential harms and will lead to reduced responsiveness to developing contexts. Another benefit to locating such bodies on terraformed planets is to ensure that colonists feel a sense of ownership over processes that legislate their ways of living and working, thus helping to ensure a sense of autonomy and determination over the development of their communities.

Another key aspect of policy should involve methods to limit the emergence of scenarios based on catastrophe, which would have repercussions for modes of governance and normative international relations, such as those depicted in Project Hail Mary. In effect this could involve linking terraforming policy to climate change policy, the detection and handling of threats to Earth's planetary integrity that potential exogenous shocks could incur and to broader policies for international conflict and relations. If, as sf has suggested, terraforming and climate change are distorted mirror-images of one another, then policy and frameworks for addressing climate change on Earth could function as the foundation for new policies and frameworks for international relations that could be built upon to co-ordinate future terraforming efforts. This is not to suggest that specific approaches to geoengineering Earth, such as through solar radiation management via stratospheric particle injection, should be adopted. Rather, the forms that international co-ordination might take could form the basis for global approaches to terraforming.

Sf also recommends policies that enable colonists to engage in deliberation, self-determination and the ability to collectively decide how terraforming should be conducted. Such deliberation and democratic processes must include communities, public, and private actors. To facilitate deliberation independent actors representing those on celestial bodies could be established, with its membership drawn from the multiple groups working on the planet, along with representatives with stakes involved in terraforming. As independence is a key source of conflict described in the literature the institution of such processes could help to ameliorate the potential for the emergence of more explosive conflict. Such deliberation need not await the terraforming of any given celestial body but could be developed on Earth through international co-ordination. Such deliberation should be a continuous process, and mechanisms should be instituted to canvass the population to ensure that the expectations relating to any form of terraforming endeavour are evaluated and re-aligned if necessary.

Access to terraformed worlds is likely to be a central issue. As Robinson's (1996a, 1996b, 1996c) Mars and Kowal's (2013, 2019a, 2019b, 2020) Lady Astronaut trilogies suggest, colonists should be selected from as diverse a pool as possible across a range of demographic dimensions. Selections should not be representative of public or private investment, and it should not be representative of the global population. Rather, diversity of opinion should be sought and opportunities to develop new collectivities embracing a range of experience, perspectives and values should be embraced. Given the long timescale involved in terraforming, experiments in fostering the emergence of such collectivities can be conducted now, and policies put in place to support such developments.

As the terraforming literature suggests, policies for re-investment into interplanetary colonies should also be instituted such that the profits and raw materials obtained through extraction on planetary bodies or asteroids do not only return to Earth. Furthermore, policies to set strict limits on the modes and scale of extraction should be factored in, and approaches that would result in the deterioration of celestial bodies prohibited. While this is important for bodies such as the Moon, where colonies and extraction could be observed from Earth, such policies should be considered for all celestial bodies. Measures should also be taken to ensure that extraction does not interfere with habitation and scientific study. In principle this means that uses of celestial bodies that would have little to no impact on its integrity should be prioritised while more transformative uses should undergo due deliberation before they are initiated, and only after all other uses of space have been satisfied.

5 Challenges to Projects of Planetary Dimension

This short chapter on terraforming policy and governance has identified several key themes evident in the sf literature about terraforming that can inform how terraforming projects are conceived and developed, or which suggest reasons for avoiding such endeavours. There is a wealth of potential models for policy and governance within the literature that highlight potential sites of conflict that could trouble future terraforming projects. Speculation and thought about terraforming scenarios can also function to help us think through the challenges of projects of planetary dimension, such as national and international responses to the effects of climate change, and to offer analyses and critiques of geoengineering, itself a form of terraforming restricted to Earth. There has been little space in this chapter to address issues of environmental ethics and the ethical responsibilities due to non-human life other than to acknowledge, in the light of Weir's (2021) Project Hail Mary and Robinson's (1996a, 1996b, 1996c) Mars trilogy, that such ethical consideration tends to be de-prioritised under the exceptional modes of governance depicted in sf that takes precedence in crisis situations. Nonetheless, these two themes take on additional resonance in the context of geoengineering where—unlike on other planets within the solar system—indigenous non-human life cannot be bracketed out of policy considerations.

The sf literature positions the success or failure of any given terraforming endeavour less in technological capacity and more as rooted in the coherence of an identity that embraces the colonists and which enables communities to work together to ensure their goals are met. Robinson's (1996a, 1996b, 1996c) Mars trilogy, in particular sequences in Blue Mars, portray deliberation as a fraught, lengthy and time-consuming process, but it also shows how such activity is crucial to ensuring that dissent and conflict do not threaten the longevity and viability of terraforming. Key to this process is the work of individuals who facilitate democratic processes. Robinson's (1996a, 1996b, 1996c) Mars trilogy emphasises the development of mechanisms to foster the work that such facilitators bring to developing diplomatic approaches to address the problems that terraforming raises. Rather than being appointed by actors on Earth, the Mars trilogy shows how such facilitators are key members of the developing communities on Mars. Trust is essential and thus facilitators must be able to foster the dialogues that emerge from each community, distil their essential elements and concerns and to disseminate those concerns to other communities. Blue Mars devotes space to depictions of such political co-ordination, which is used to bring actors and communities together to debate key issues for the given celestial body.

The failure to account for the goals and motivations of the colonists is well-attested in sf as a source of conflict and dissent which must be overcome before the success of terraforming projects can be ensured. Often, as in Killick's (2021) Deimos, Weir's (2021) Artemis or McDonald's (2015, 2017, 2019) Luna trilogy, those whose interests have been typically marginalised and those who leave Earth to find new opportunities for work and the development of fulfilling lives become key players in the process of opening up spaces where meaningful choices about how terraforming should be organised and conducted can be made. As the more cynical of these stories suggest, opportunities for deliberation and self-determination are not compatible with the goal of establishing an economically viable colony or are not possible within the context of the systems of governance under which such colonies operate. Yet even in such stories dissenting voices do transform aspects of those systems and integrate themselves into the new frameworks so instantiated.

References

Aldiss, B., & Penrose, R. (1999). *White Mars; Or, the mind set free, a 21st century utopia.* Little, Brown.

Attebery, B., & Hollinger, V. (Eds.). (2013). *Parabolas of science fiction.* Wesleyan University Press.

Boyer, D., & Szeman, I. (2014, February 12). *The rise of energy humanities: Breaking the impasse.* University Affairs. www.universityaffairs.ca/opinion/in-my-opinion/the-rise-of-energy-humanities/

Broderick, D. (1995). *Reading by starlight: Postmodern science fiction.* Routledge.

Butler, A. (2014). Futurology. In R. Latham (Ed.), *The Oxford handbook of science fiction* (pp. 513–512). Oxford University Press.

Cater, C. I. (2014). Steps to space: Opportunities for astrotourism. *Tourism Management, 31*(6), 838–845.

Heinlein, R. A. (2001). *The moon is a harsh mistress.* Gollancz.

Killick, J. (2021). *In the shadow of Deimos: A terraforming Mars novel.* Aconyte.

Kowal, M. R. (2013). *The lady astronaut of Mars.* Tor. https://www.tor.com/2013/09/11/the-lady-astronaut-of-mars/

Kowal, M. R. (2019a). *The calculating stars.* Solaris.

Kowal, M. R. (2019b). *The fated sky.* Solaris.

Kowal, M. R. (2020). *The relentless moon.* Solaris.

McDonald, I. (2015). *Luna: New moon.* Gollancz.

McDonald, I. (2017). *Luna: Wolf moon.* Gollancz.

McDonald, I. (2019). *Luna: Moon rising.* Gollancz.

Milner, A. (2012). *Locating science fiction.* Liverpool University Press.

Milner, A., & Burgmann, J. R. (2020). *Science fiction & climate change: A sociological approach.* Liverpool University Press.

Morrissey, T. J. (2000). Ready or not, here we come: Metaphors of the Martian megatext from Wells to Robinson. *Journal of the Fantastic in the Arts, 10*(4), 372–394.

Pohl, F., & Kornbluth, C. M. (1974). *The space merchants.* Random House.

Robinson, K. S. (1996a). *Blue Mars.* Voyager.

Robinson, K. S. (1996b). *Green Mars.* Voyager.

Robinson, K. S. (1996c). *Red Mars.* Voyager.

Robinson, K. S. (2018). *Red moon.* Orbit.
Seed, D. (2000). Science fiction dialogues. In A. Sawyer & D. Seed (Eds.), *Speaking science fiction: Dialogues and interpretations* (pp. 11–20). Liverpool University Press.
Steinmüller, K. (2003). The uses and abuses of science fiction. *Interdisciplinary Science Reviews, 28*(3), 175–178.
Weir, A. (2017). *Artemis*. Penguin.
Weir, A. (2021). *Project hail Mary.* Penguin.
Williamson, J. (2004). Collision orbit. In *Seventy-five: The diamond anniversary of a science fiction pioneer* (pp. 216–277). Haf.

FSC
www.fsc.org
MIX
Papier aus verantwortungsvollen Quellen
Paper from responsible sources
FSC® C105338

If you have any concerns about our products,
you can contact us on
ProductSafety@springernature.com

In case Publisher is established outside the EU,
the EU authorized representative is:
Springer Nature Customer Service Center GmbH
Europaplatz 3, 69115 Heidelberg, Germany

Printed by Libri Plureos GmbH
in Hamburg, Germany